ICE PRESENTS

UN 75

SUSTAINABLE ENGINEERING IN ACTION

Printed in the UK by Pureprint Limited on 130gsm Amadeus Silk. This paper has been independently certifieds according to the standards of the Forest Stewardship Council® (FSC)®.

ISBN: 978-1-911339-43-4

This year marks the 75th anniversary of the founding of the United Nations, and this publication aims to highlight how engineering and those sectors associated with the built environment are tackling many of the most pressing challenges of the day, as represented by the UN's Sustainable Development Goals.

The breadth of projects and organisations profiled in this book illustrates the extent of the challenge that the SDGs represent, along with the range of responses to that challenge. From fighting poverty, hunger and inequality; to promoting health, peace, education and economic development; to protecting the environment, the subjects addressed herein reflect what engineers have always done – take on the world's big problems with solutions large and small.

As much as any of the SDGs, the last of the UN's goals – SDG 17: Strengthen the means of implementation and revitalise the global partnership for sustainable development – is key to the ICE's principles. Our industry is made all the better and more effective by the sharing of its knowledge and expertise, as this book attests.

Foreword

I am delighted to introduce the latest edition of our partnership books series with publisher St James House, which examines how our profession has increased its emphasis on sustainable engineering solutions.

Since the UN was founded 75 years ago, many things have changed. The civil engineering profession has moved from concentrating on post-war rebuilding to the modern day practice: drawing on sensors, AI and data analytics underpinned with an emphasis on how infrastructure delivers social value while also mitigating the effects of climate change and removing carbon from the atmosphere. But the fundamentals are the same: civil engineers shape the way the world is, now, and in the future. We exist to create a better, enduringly sustainable, world for our children and grandchildren.

This book sets out how civil engineers change lives on a truly global scale. With skills, expertise, judgement and recognising that infrastructure often lasts for many lifetimes, engineers have learned to build adaptable, resilient and carbon-neutral solutions to some of the biggest challenges we face across the world. If we are truly to change our world for the better we will need to take a long-term, system of systems and carbon-based approach.

Civil engineers are often called the invisible superheroes, as their work regularly goes unnoticed. But it has huge beneficial long-term effects for the whole of humankind. The challenge, as for many professions, is to entice the next generation of superheroes into this fascinating work. We need to get young people to realise this as a creative, enjoyable and hugely rewarding career, where they should want to spend their working lives.

Civil engineers' interest in using the UN's SDGs has grown over time, and this book shows the current level of sophistication in achieving sustainability. I encourage you to read it, learn from it the art of the possible, and reflect on how much better we can become. In 75 years, in 2095, I hope that my successor will look back at what we did today and decide that we lived up to the challenge.

Rachel Skinner
ICE President

Foreword

Infrastructure is key to sustainable development – from homes and hospitals, to roads and schools, to water and waste management, to digital communications, infrastructure provides the backbone of a functioning society.

Research has shown that infrastructure influences the achievement of 92 per cent of all targets across the 17 Sustainable Development Goals (SDGs). Importantly, infrastructure influences development far into the future, both positively and negatively. So it is crucial to make the right infrastructure decisions now, if we want to reduce inequality, maintain a healthy population and protect our environment over the long term.

Our world's infrastructure needs are huge. It is estimated that the world needs over $90 trillion in global infrastructure investment to support sustainable development by 2040.

Globally, more than 2 billion people lack access to safe drinking water and 4.5 billion lack access to sanitation, while 975 million people do not have access to electricity and 1 billion lack access to all-weather roads. Unsurprisingly, the burden of poor infrastructure falls disproportionately on vulnerable and marginalised groups, including women and girls, and disabled people. The scale and complexity of this need makes an even stronger case for our collective responsibility to make the right decisions about our infrastructure systems.

The COVID-19 crisis has further highlighted our world's massive infrastructure gaps. It has exposed how fundamentally unprepared our world's infrastructure was to deal with a crisis of this scale. And it continues to spread human suffering and destabilise the global economy, with consequences that will be with us for years to come.

But COVID-19 has also shown how interconnected our challenges are, and that we can only overcome them if we work together. That is the essence of the SDGs – to achieve them, we need to mobilise all our resources, across borders, generations and sectors.

This is all the more relevant now, as we mark 75 years since the formation of the United Nations.

As we emerge from the COVID-19 crisis, it is important to recover in a way that puts us on the right path to achieve the SDGs. Focusing on quality infrastructure is a key part of this journey. As we recover, our infrastructure needs to be more inclusive, sustainable and resilient, so that it can adapt and scale in the face of crises, from pandemics to the impact of a changing climate.

This book, published by the Institution of Civil Engineers, provides great insight into how we can work together to improve lives, by ensuring that our world gets the infrastructure systems it needs to thrive. Achieving the SDGs is everyone's business, and this publication is a timely reminder of this important issue.

Grete Faremo
Under-Secretary-General and UNOPS Executive Director

Contents

Education, rights and equality

Introduction

The UN was established at the end of the Second World War to promote international cooperation and peace. At the time, that was a pretty audacious goal to strive for, but there is something to be said for setting a big aim, and how this engages and energises people to achieve it. So far there hasn't been a Third World War, so we can claim success on this target. The corollary benefits of such multinational cooperation – seen in post-war reconstruction, the globalised economy, research, academia, industry and almost every profession – is incalculable.

Just one example of such collaboration is the UN's Sustainable Development Goals (SDGs). The 17 SDGs were set in 2015 and billed as the shared blueprint for peace and prosperity for people and the planet. They aim to end poverty, improve health and education, reduce inequality, and spur economic growth, all while tackling climate change.

The rate of development since the SDGs were agreed has been slower than ideal, but inertia is a powerful force to counteract. With these goals to live by and strive for, we can hope for a better, more efficient, fairer, more stable world, with fewer harmful effects from climate change; surely an ideal we can all get behind.

Recent developments in engineering, such as the increasing understanding of BIM, data and analytics, and carbon tracking and AI, show that as a profession we can develop and learn. The good news is the well of human creativity springs eternal, and with these powerful modern tools we can make inroads into the great challenges that we face. In fact, without these sorts of tools and development, we face no chance.

Social and environmental considerations are playing an ever-more significant part in the education of engineers, as evidenced in one of the very articles within this book ("Strength in breadth", pp. 146–47). Civil engineering has gradually widened as a professional role since its origin in the Victorian age, and the modern civil engineer does far more than a 19th-century engineer would even recognise. This broadening is beneficial, as it brings with it more ways to view challenges. But we need this broader, enlightened view to be more successful at solving problems, be they engineering, societal, environmental, economic, climatic or wider.

Recently, from various disasters we have seen some failures to apply this "systems thinking" overview, and each time we rightly say "no more, we must learn and do better". This applies just as much to new building as to maintenance of infrastructure; the whole lifecycle view must be used to point out that short-term savings in minimised maintenance do not come without a cost, and may be detrimental overall. Taking the broader view of problems helps elucidate this and therefore we encourage it, so that in the long run we will benefit from better running and less risky infrastructure.

The core purpose of ICE has remained unchanged since its founders laid down its original resolutions in 1818. Namely, to support and advise the profession and promote the sharing of knowledge to benefit the wider society and advance scientific development. The Institution does this through a variety of actions, from public strategy to individual support.

ICE plays an active and enthusiastic role in policy and public affairs, with strong links with the worlds of business, academia, think tanks and finance, to ensure that ICE members' voices are represented at the highest level of national and international political and commercial decision-making. An example of this is the Enabling Better Infrastructure programme, which offers insights to governments and decision-makers on how to best plan and deliver infrastructure with the most effective social and economic outcomes. More recently, the Infrastructure Client Group-led industry-wide-change programme Project 13 aims to improve infrastructure, deliver a better workforce and create a more sustainable, productive construction industry.

"Our mission is to make the SDGs accessible to built environment professionals, so they can ensure that infrastructure enables sustainability, rather than diminishes it"

In 2000, ICE established CEEQUAL as evidence-based sustainability assessment awards for civil engineering, infrastructure, landscaping and public-realm projects. Since then, it has helped thousands of projects celebrate the achievement of high environmental and social performance, and, crucially, gain credibility and plaudits, in the hope that this changes the industry to become the norm. It was a world-first and remains the leading international infrastructure rating scheme.

In 2018, ICE hosted the Global Engineering Congress, bringing thousands of delegates from around the world to discuss the SDGs. The outcome was a commitment to the goals and a route map to progress this intent into actions. The Sustainability Route Map outlines our activities over the next three years. Our mission is to make the SDGs accessible to built environment professionals, so they can ensure that the infrastructure they build enables sustainability, rather than diminishes it.

Knowledge sharing is a key activity of ICE, and as such we are particularly proud of our resource library. This had an auspicious origin, starting from the donation of Thomas Telford's books and papers 200 years ago. The stunning ICE headquarters house the main physical library at One Great George Street in London and a dedicated team looks after member queries. ICE has been a publisher since 1936 and now provides a vast digital resource of journals with its virtual library – the largest collection of civil engineering resources in the world – digitised and made available to members online. ICE also provides contractual services to the construction sector through the NEC (New Engineering Contract) family of contracts.

All of this goes to show the effort and time taken to develop better infrastructure. This is a never-ending task, but one that offers great rewards, and a greatly rewarding profession for those who choose to become a civil engineer.

I thank all the contributors, firstly for their work and secondly for recording and sharing their thoughts within this book; knowledge is built in this sort of selfless sharing. All activity towards greater sustainability is to be commended, so thanks also to you the reader; I hope this provides some food for thought as well as world-leading examples of sustainability in action.

Mark Hansford
ICE Director of Engineering Knowledge

TRANSFORMING THE WORLD

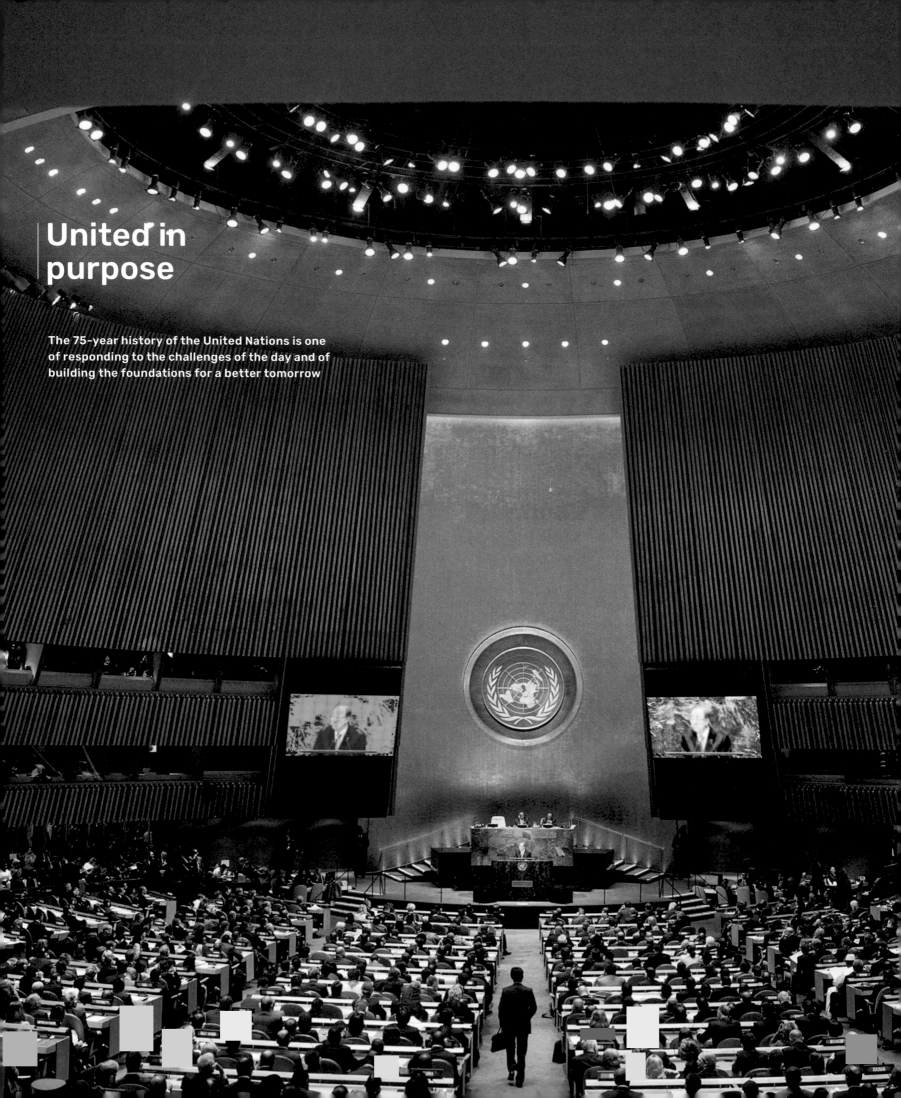

United in purpose

The 75-year history of the United Nations is one of responding to the challenges of the day and of building the foundations for a better tomorrow

The United Nations is a global organisation dedicated to maintaining international peace and stability, developing friendly relations among nations and promoting social progress, better living standards and human rights. It was officially established in 1945 following the Second World War, initially with 51 member states. Today, the organisation, which is headquartered in New York City, has 193 members, encompassing nearly every country on earth.

Although best known for peacekeeping, peacebuilding, conflict prevention and humanitarian assistance, there are many other ways the UN – along with its specialised agencies, funds and programmes – affects our lives and makes the world a better place. The UN has negotiated 172 peace settlements that have ended regional conflicts and is credited with participation in over 300 international treaties on topics as varied as human rights conventions to agreements on the use of outer space and the oceans.

It works on a broad range of fundamental issues, from sustainable development, environment and refugees protection, disaster relief, counter terrorism, disarmament and non-proliferation, to promoting democracy, human rights, gender equality and the advancement of women, governance, economic and social development, international health, clearing landmines, expanding food production, and more. All of these are carried out in order to achieve its goals and coordinate efforts for a safer world for this and future generations.

The UN was officially founded in 1945, but its roots lie in several meetings conducted throughout the Second World War. On 12 June 1941, St James's Palace in London was the home of an inter-allied declaration from five Commonwealth nations (the UK, Canada, Australia, New Zealand and South Africa) along with nine European governments that were then exiled in London (France, Belgium, Czechoslovakia, Greece, Luxembourg, the Netherlands, Norway, Poland and Yugoslavia). Even though the Axis powers were in the ascendancy at this point, these nations were determined to not only win the war but to look beyond victory.

"Would we win only to live in dread of yet another war?" they announced. "Should we not define some purpose more creative than military victory? Is it not possible to shape a better life for all countries and peoples and cut the causes of war at their roots?" The 14 nations came to an agreement that would provide the basis for the United Nations. "The only true basis of enduring peace is the willing cooperation of free peoples in a world in which, relieved of the menace of aggression, all may enjoy economic and social security," read the agreement. "It is our intention to work together, and with other free peoples, both in war and peace, to this end."

ORIGINS OF THE UN

Two months later, on 14 August, British Prime Minister Winston Churchill met US President Franklin D Roosevelt on an undisclosed ship in the mid-Atlantic to sign what became known as the Atlantic Charter. Even though the US would not enter the war for another four months, this charter declared certain common principles in the two countries' national policies "on which they based their hopes for a better future for the world". These included "safety within national boundaries, freedom from fear and want". The Atlantic Charter provided the basis for what became known as the United Nations Declaration, which was signed on New Year's Day 1942 by Roosevelt and Churchill, along with Maxim Litvinov of the USSR and TV Soong of China. It was the first time that the term "United Nations" had been used in the context of this organisation.

"Being convinced that complete victory over their enemies is essential to defend life, liberty, independence and religious freedom, and to preserve human rights and justice in their own lands as well as in other lands, and that they are now engaged in a common struggle against savage and brutal forces seeking to subjugate the world," read the declaration. The document was signed by 26 Allied nations, including India, Cuba, El Salvador, Haiti and Panama, and throughout 1942 was signed by another 21, including Mexico, Ethiopia, Iran, Iraq and Brazil.

This was followed by two conferences among the key Allied powers in 1943. On 30 October that year, the US Secretary of State Cordell Hull flew to the Soviet Union to sign the Moscow Declaration, alongside British Foreign Secretary Anthony Eden, Soviet Foreign Minister Vyacheslav Molotov and China's Ambassador to the Soviet Union Foo Ping Shen. All four agreed to "recognise the necessity of establishing at the earliest practicable date a general international organisation, based on the

principle of the sovereign equality of all peace-loving states, and open to membership by all such states, large and small, for the maintenance of international peace and security."

On 1 December the three main Allied leaders – Roosevelt, Stalin and Churchill – met in the Iranian capital Tehran. "We are sure that our concord will win an enduring peace," they confirmed. "We recognise fully the supreme responsibility resting upon us and all the United Nations to make a peace which will command the goodwill of the overwhelming mass of the peoples of the world and banish the scourge and terror of war for many generations."

THE CHARTER IS AGREED

On the 7 October 1944 came another meeting between representatives of the UK, the US, the USSR and China, this time at Dumbarton Oaks in Washington DC, outlining a proposal for a world organisation. This quite detailed plan outlined a proposed structure for the UN. A further meeting between Stalin, Churchill and Roosevelt at the Crimean resort of Yalta on 11 February 1945 finalised voting procedures for the organisation. It was followed by a conference in San Francisco in April and June 1945, where representatives of all 50 nations who had signed the 1942 document completed the Charter of the United Nations.

By this time, the structure of the organisation had been agreed. In addition to the General Assembly of all member states (the UN's "town hall") and a Security Council of five permanent and 10 non-permanent members, the Charter provided for an Economic and Social Council, an International Court of Justice, and a Trusteeship Council to oversee certain colonial territories to independence (something that was only made obsolete in 1994 when Palau was granted independence from the United States). All of this was under the executive power of the Secretariat, led by a Secretary-General.

The UN Charter outlined the four key principles of the UN: to maintain international peace and security; to develop friendly relations among nations; to achieve international cooperation in solving international problems; and to be a centre for harmonising the actions of nations in the attainment of these common ends. The UN officially came into existence on 24 October 1945, when representatives of the 50 nations ratified this charter. The first meeting of the General Assembly followed in London on 17 January 1946, when Prime Minister Clement Attlee made the opening address at Central Hall, Westminster.

The UN was not, of course, the first, or even the second, international organisation to be formed after a devastating war. The Concert of Europe was formed at the Congress of Vienna in 1814 and 1815 after the Napoleonic wars to maintain the status quo between European powers and thus avoid war. It broke down by the 1850s, but it did pave the way for the establishment of the first Geneva Conventions and the Red Cross.

BEFORE THE UN

However, the most important precedent for the UN was the League Of Nations, created at the 1919 Paris Peace Conference after the devastation of the Great War. Like the UN, it was devoted to the settling of disputes and the prevention of war, and it had some limited successes – settling border disputes between Finland and Sweden, between Albania, Greece and Yugoslavia, and between Greece and Bulgaria. But it also stood by as Japan invaded China in 1931, as the Italians invaded Ethiopia in 1935, as Germany and Italy intervened in the Spanish Civil War in 1933, and as the Germans reoccupied Rhineland in 1936. It also did not give any representation to those living under colonial powers.

Part of the problem for the League of Nations was that its founding "Covenant" was tied to the peace treaty with Germany, and the link between the two was always problematic, both for the defeated Central Powers (many Germans regarded the Treaty of Versailles as an affront) and even for victors such as the United States (despite President Woodrow Wilson's key role in its establishment, the US Senate never ratified the Covenant and the US never joined the League of Nations). America's lack of involvement was just one key absence – the Soviet Union didn't join the League until 1934, around the same time that Germany, Italy and Japan withdrew from the League to pursue their expansionist aims.

From the start, the creators of the UN established several key differences with the League to avoid these problems. It was crucial that the UN Charter was the product of the combined efforts of 50 nations at the 1945 San Francisco Conference, rather than just the elite Security Council, as it ensured that the views of smaller and

developing nations were taken into account. It was also important, however, to avoid the trap caused by the League of Nations' Covenant, whereby decisions had to be unanimous. This gave any sovereign nation the power to veto any resolution, rendering the League largely powerless. In the UN, only members of the permanent five members of the Security Council – the UK, the US, France, China and the USSR – can veto a measure, and each has regularly used this power to protect either itself or its allies. In essence, the United Nations was created with the belief that only a multilateral institution could guarantee world peace.

Of course, the presence of both the Soviet Union and the United States on the UN Security Council interfered with the UN's basic peacekeeping mission throughout the Cold War. It hampered the organisation's ability to take action in any situation where the interests of either of these two superpowers were involved, or where they were committed to opposing sides

in disputes involving smaller nations. The one Cold War conflict that the UN did intervene in came in 1950, when North Korea invaded South Korea. The Soviet Union was boycotting the UN at the time, so the UN Security Council voted in favour of stemming North Korea's aggression. This ended with an armistice in 1953, with a demilitarised zone that divides the peninsula to this day. But, as the Soviet Union re-entered the UN, the organisation was unable to check the Soviet invasion of Hungary in 1956 or Czechoslovakia in 1968.

AN EXPANDING MEMBERSHIP
Throughout the 1950s, '60s and '70s, decolonisation transformed the composition and functioning of the UN, with dozens of African and Asian countries gaining independence from France, Britain, Portugal, Spain, Belgium and the Netherlands. By 1961, the UN's membership had more than doubled to 104; by 1979 there were 152

members. These newly independent countries sought to redirect the energies of the organisation toward easing the transition to independence. The UN deployed a peacekeeping force to police an end to the Suez crisis in 1956, while UN Emergency Forces were deployed to keep the peace on the Egypt/Israel border (from 1956 to 1967 and then from 1973 to 1979), in the Congo (from 1960 to 1964), in Cyprus (from 1964) and in the Lebanon (from 1978). It also agreed to a voluntary arms embargo against apartheid South Africa in 1963 (which became mandatory in 1977), agreed to sanctions against Rhodesia in 1966, and in 1969 declared South Africa's presence in Namibia illegal and assumed responsibility for the territory.

As a result of the activism of states referred to during the Cold War as the "Third World", the UN took on additional responsibilities for economic, political and social development, and the monitoring and enforcement of fundamental human rights. Its role expanded to include refugees, environment, weapons, health and even global criminal justice. This meant that much of the UN's energy was devoted to assisting in the economic, political and social development of these new nations, and in monitoring and enforcing fundamental human rights in the developing world. In 1954, the UN High Commissioner for Refugees won the first of two Nobel Peace Prizes (it won another in 1981); in 1965, UNICEF also won a Nobel Peace Prize; as did the UN's International Labour Organisation in 1969.

The Cold War stand-off that curbed the UN's powers seemed to come to an end when Mikhail Gorbachev became Soviet leader in 1985. Unlike his predecessors, he called on the UN to play a more central role in world politics as a cornerstone of global security, something no previous Soviet leader had ever done. It ushered in a new age of consensus among the UN's Security Council. From 1988 to 1993, 20 UN military operations were launched, more than during the entire first four decades of the organisation, earning the UN a Nobel Peace Prize.

The response to Iraq's invasion of Kuwait in August 1990 saw the UN Security Council unanimously backing "all means

BELOW
Gorbachev addresses
the UN General Assembly
in 1988

OPPOSITE
UN peacekeepers on the
border between Lebanon
and Israel, November 2009

necessary" to eject Saddam Hussain's forces, leading to the Gulf War in February 1991. Other international successes followed. The UN oversaw the independence of Eritrea in 1993, supervised elections in Cambodia in 1993 and provided observers for South Africa's first universal elections in April 1994. Peacekeeping actions in Angola (1991), Sierra Leone (1999), Timor-Leste (1999) and Kosovo (1999) showed that the post-Cold War world was becoming more willing to work through the UN in the pursuit of peace and international security.

After Cote d'Ivoire's civil war in 2004, the UN deployed 6,000 peacekeepers in 2006, increasing the number to 12,000 in 2011. It

successfully disarmed 70,000 combatants and re-integrated them into society and oversaw the return of more than a quarter of a million refugees. The UN has also overseen successful peacekeeping operations in El Salvador, Guatemala, Mozambique, Namibia, Tajikistan, Liberia and Haiti. It authorised military intervention in Libya in 2011, and successfully prosecuted Liberian leader Charles Taylor and Serbian President Slobodan Milosevic for war crimes. The UN has also encouraged two countries to give up nuclear weapons – Kazakhstan after the fall of the Soviet Union, and South Africa after the end of apartheid.

These recent successes have been accompanied by more troubled peacekeeping missions, such as those in Rwanda, South Sudan, Somalia and the former Yugoslavia, and the ongoing conflict in Syria. In the face of these challenges, the UN High Commissioner for Refugees has performed admirably. Created in 1950 to deal with the millions of Europeans displaced after the Second World War, it also become a huge part of the UN's function, and is still on the front lines of the struggle to feed and house millions of refugees fleeing Syria, Iraq, Afghanistan, Somalia and other countries affected by war and poverty.

A GLOBAL ROLE

When considering the UN purely through the prism of conflict, it is sometimes easy to overlook its achievements in health, education and humanitarian projects. Since 1945, it has provided food to 90 million people in over 75 countries; assisted more than 34 million refugees; and authorised 71 international peacekeeping missions. It has worked with 140 nations to minimise climate change, provided vaccinations for 58 per cent of children in the world, and helped around 30 million women a year with maternal health efforts. In 2004, it responded to the Indian Ocean tsunami with a relief fund that raised more than $6.25 billion. In the support of democracy it has assisted around 50 countries a year with their elections, and protected human rights with 80 treaties and declarations. The UN's Universal Declaration of Human Rights of 1948 is the most translated document in the world, proclaiming rights such as the right to not be enslaved, the right to free expression, and the right to seek asylum from persecution from other countries.

In 1948, the UN created the World Health Organization to deal mainly with communicable diseases like smallpox, tuberculosis and malaria. By 1980, after the organisation had led a 13-year immunisation campaign, the WHO declared smallpox extinct. This was a mammoth task that involved locating and isolating every victim of the disease, and inoculating as many people as possible in

the vicinity. In 1986, the UN announced a target to eradicate guinea worm, a disease affecting 3.5 million people – a campaign that has been largely successful. The WHO has also had success in its initiatives with HIV/Aids and viruses such as Ebola, SARS, MERS and COVID-19.

Its cultural agency, UNESCO, has also made enormous strides in education, as well preserving some of the planet's most important places through its World Heritage Sites. In 1978, the Galapagos Islands and the Taj Mahal were among 1,000 sites granted this status, something that became the international benchmark for protecting the world's most important natural and historic places. UNESCO has helped with their management, upkeep and environmental protection, and also intervened when they were under threat. When construction began in 1995 on an eight-lane motorway a mile south of the Giza pyramids and the Sphinx in Egypt, UNESCO dispatched an expert mission to Cairo to meet with government authorities about diverting the road, which Egyptian officials agreed to do.

DRIVING DEVELOPMENT

In the past half century the UN has pioneered environmental initiatives, setting up the UN Environment Programme in 1972. In 1987, it forged a treaty on the protection of the ozone layer in Montreal, where 24 countries signed a milestone accord that promised to halve the production and use of ozone-destroying chemicals by 1999. In June 1992, came the UN's first "Earth Summit" in Rio de Janeiro, where a treaty on climate change was adopted. It was followed by the Kyoto Protocol of 1997 (which legally binds industrialised nations to reduce worldwide emissions of greenhouse gases by an average of 5.2 per cent below their 1990 levels) and the 2015 Paris Agreement to fight climate change.

In 2000, the UN Millennium Development Goals (MDGs) were created to encourage progress against poverty and malnutrition, while also promoting human rights, gender equality, environmental sustainability and education. Before this, somewhat amazingly, there had never been a common framework for promoting global development, but it had become evident that one was needed. After the end of the Cold War, the Soviet Union's aid to its allied developing countries disappeared, and many wealthy Western countries cut their foreign aid budgets and turned their focus inward. Meanwhile, institutions such as the IMF and the World Bank encouraged developing countries to cut spending on public infrastructure programmes in the name of efficiency as a condition for receiving support. This had troubling results for much of the developing world.

The overarching vision of cutting the amount of extreme poverty in half by 2015 was anchored in a series of specific goals, attention and resources to many issues that might otherwise have been forgotten. The MDGs mobilised governments and business leaders to donate tens of billions of dollars to life-saving tools, such as antiretroviral drugs and modern mosquito nets. They promoted cooperation among public, private and nongovernmental

RIGHT
The Sustainable Development Goals
are projected onto the facade of the
General Assembly building in New York,
September 2015

organisations (NGOs). Bill Gates, in a 2008 address to the UN General Assembly, described the MDGs as "the best idea for focusing the world on fighting global poverty that I have ever seen".

The MDGs were uneven but were widely admired for their impact. The Brookings Institution has praised them for kickstarting progress where it was lacking, especially in Africa, where they have helped to reduce extreme poverty, increase education and provide support for small subsistence and cash-crop farmers. Moreover, it has massively increased access to healthcare, reducing child mortality, encouraged pharmaceutical companies to make medicine more widely available, and addressed the HIV/Aids epidemic across Africa.

ESTABLISHING THE SDGs

Following the MDGs came the Sustainable Development Goals (SDGs). The UN conducted the largest consultation programme in its history to gauge opinion on what these goals should include and, in 2015, the General Assembly unanimously endorsed this 17-item list of objectives to achieve before 2030, which includes addressing climate change, pollution, poverty and hunger; and eliminating gender discrimination.

There was some concern from governments and NGOs that there were too many goals, but there was also a general consensus that it was better to have 17 goals that include targets on women's empowerment, good governance, and peace and security, for example, than fewer goals that didn't address these issues. To achieve these goals, the UN, together with the governments of 177 countries and the support of the private sector, has set up a fund, called the United Nations Development Programme (UNDP).

As it celebrates its 75th anniversary, the UN still faces challenges. It has done much to address allegations of bureaucratic inefficiency and waste, and has made an effort to enhance transparency – when Ban Ki-Moon stepped down in 2016, the UN held its first public debate between the candidates for Secretary-General.

Portuguese diplomat António Guterres, who took over as the ninth Secretary-General of the UN in 2017, highlighted several key goals for his administration, including an emphasis on diplomacy for preventing conflicts, more effective peacekeeping efforts, and streamlining the organisation to be more responsive and versatile to global needs. "With the structural aspects of the reforms now well consolidated," said Guterres, "it is imperative to keep the foot in the pedal to achieve the cultural change we need for greater collaboration across pillars and tangible results for people on the ground."

It's a mandate that has garnered international recognition with, for instance, the UN's World Food Programme winning this year's Nobel Peace Prize. It is also one that has sustainable development and the efforts and ingenuity of engineers around the world at its heart.

Blueprint for a better world

The UN's Sustainable Development Goals have set the agenda for individuals and organisations around the world, and galvanised governments into action for the next decade

"We don't have a Plan B because we don't have a Planet B," said the then UN Secretary-General Ban Ki-moon in a 2016 press conference. "We have to work very hard, very seriously and urgently. This is what I am saying, not as only Secretary-General, but as one of the citizens of this world. I think you and I, and all of us have a common moral responsibility."

He was talking at an event in Marrakech at the Sustainable Innovation Forum, addressing climate change, but his words have become a rallying cry in the development of the United Nations' Sustainable Development Goals (SDGs). These might be the most important set of targets ever issued – 17 goals that set about ending poverty, protecting the planet and improving the lives and prospects of the entire human population. They involve a combination of economic growth, social inclusion and environmental protection.

The 17 goals, listed overleaf, were adopted by all UN member states in 2015, as part of the 2030 Agenda for Sustainable Development, which set out a 15-year plan to achieve them. The final list was created from 169 targets, with each SDG having around 10 targets and coming with a series of indicators to measure progress towards these being met.

One key feature is the interconnectedness of the goals. The SDGs recognise that ending poverty and other deprivations must go hand-in-hand with strategies that improve health and education, reduce inequality and spur economic growth – all while tackling climate change and working to preserve our oceans and forests. As a result, many of the goals overlap. Issues such as women's rights, education and democracy cut across several SDGs; as does addressing poverty, starvation and climate change. The roles of scientists and civil engineers are as important as government policy makers.

The SDGs build on decades of work by the UN. In June 1992 at the "Earth Summit" in Rio, more than 178 countries adopted Agenda 21, a comprehensive plan of action to build a global partnership for sustainable development to improve human lives and protect the environment. UN member states unanimously adopted the Millennium Declaration at the Millennium Summit in New York, in September 2000, which led to the Millennium Development Goals (MDGs), a United Nations initiative that issued eight goals to be achieved by 2015. These were: 1) to eradicate extreme poverty and hunger, 2) to achieve universal primary education, 3) to promote gender equality and empower women, 4) to reduce child mortality, 5) to improve maternal health, 6)

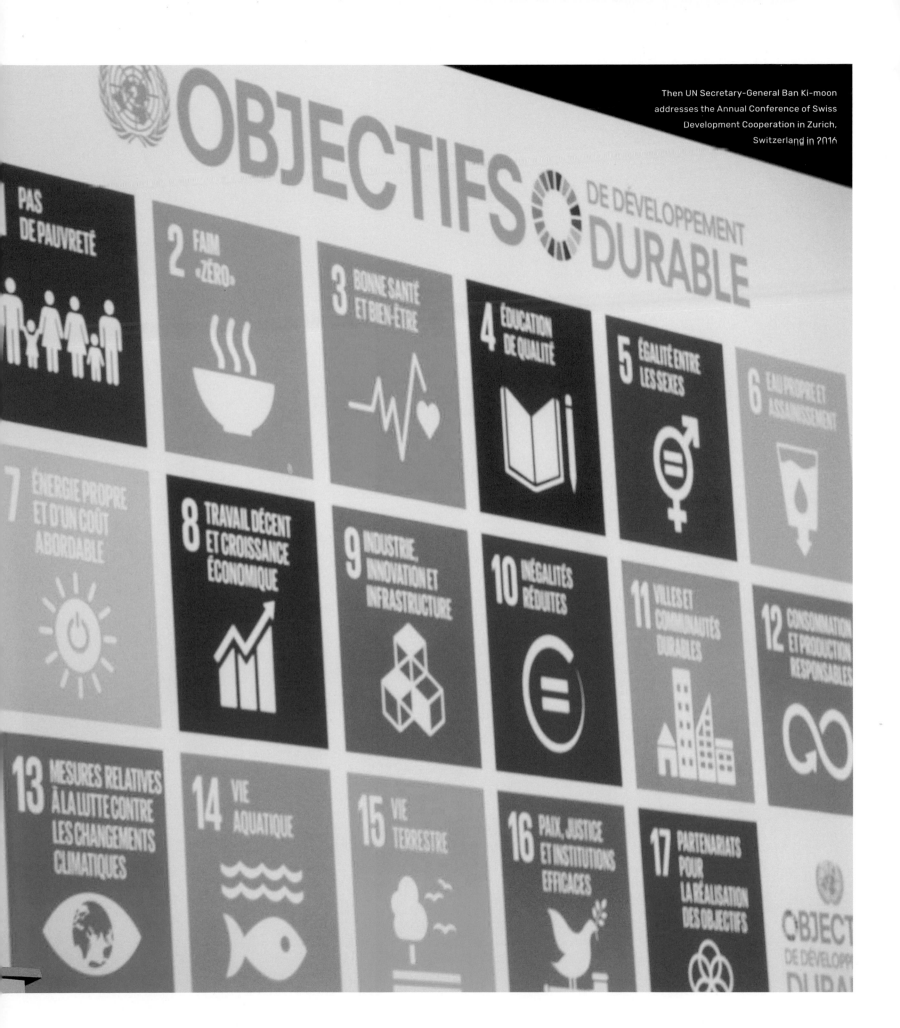

> *"We envisage a world free of poverty, hunger, disease, and want, where all life can thrive"*

to combat HIV/AIDS, malaria and other diseases, 7) to ensure environmental sustainability and 8) to develop a global partnership for development.

The MDGs did make great progress: initiating developments to reduce poverty, increase access to education and to support farmers. Many G8 economies agreed to cancel the debt of developing nations, and huge strides were made to address diseases, particularly the AIDS pandemic that swept across much of Africa. However, the UN was aware of the MDGs' drawbacks. In particular, MDG 8 ("to develop a global partnership for development") was interpreted as indicating a "donor–recipient" relationship between "first world" and "third world" countries.

To avoid this, a key aim of its follow-up would be to favour collective action by all countries. The aim was also to cover more ground and address the root causes of poverty, with ambitions to address inequalities, economic growth, decent jobs, city planning, industrialisation, oceans, ecosystems, energy, climate change, sustainability, peace and justice.

While establishing the SDGs, the UN conducted the largest consultation programme in its history. An "Open Working Group" was established in 2013 to decide on its methods of work, to ensure the full involvement of relevant stakeholders, including expertise from civil society, scientists and Indigenous peoples. Together they came up with 169 targets upon which the SDGs were based.

In January 2015, the General Assembly began the negotiation process on the post-2015 development agenda. The process culminated in the subsequent adoption of the 2030 Agenda for Sustainable Development, with 17 SDGs at its core, at the UN Sustainable Development Summit in September 2015. It coincided with some other groundbreaking agreements of that year, including the Sendai Framework for Disaster Risk Reduction (March 2015),

the Addis Ababa Action Agenda on Financing for Development (July 2015) and the Paris Agreement on Climate Change (December 2015).

With a little over 10 years left to achieve the Sustainable Development Goals, world leaders at the SDG Summit in September 2019 called for a Decade of Action and delivery for sustainable development, and pledged to mobilise financing, enhance national implementation and strengthen institutions to achieve the goals by the target date of 2030.

The UN Secretary-General, António Guterres, called on all sectors of society to mobilise for a decade of action on three levels: global action to secure greater leadership, more resources and smarter solutions for the SDGs; local action embedding the needed transitions in the policies, budgets, institutions and regulatory frameworks of governments, cities and local authorities; and people action, including by youth, civil society, the media, the private sector, unions, academia and other stakeholders, to generate an unstoppable movement pushing for the required transformations.

"We are setting out a supremely ambitious and transformational vision," says the UN's 2020 SDG report. "We envisage a world free of poverty, hunger, disease and want, where all life can thrive. We envisage a world free of fear and violence. A world with universal literacy. A world with equitable and universal access to quality education at all levels, to health care and social protection, where physical, mental and social well-being are assured. A world where we reaffirm our commitments regarding the human right to safe drinking water and sanitation and where there is improved hygiene; and where food is sufficient, safe, affordable and nutritious. A world where human habitats are safe, resilient and sustainable and where there is universal access to affordable, reliable and sustainable energy."

For a full list of the SDGs and their targets, go to page 248.

The Sustainable Development Goals

SGD 1
End poverty in all forms everywhere

SDG 2
End hunger, achieve food security and improved nutrition and promote sustainable agriculture

SDG 3
Ensure healthy lives and promote well-being for all at all ages

SDG 4
Ensure inclusive and equitable quality education and promote lifetime opportunities for all

SDG 5
Achieve gender equality and empower all women and girls

SDG 6
The UN's goal to "ensure availability and sustainable management of water and sanitation for all" in depth

SDG 7
Ensure access to affordable, reliable, sustainable and modern energy for all

SDG 8
Promote sustained, inclusive and sustainable economic growth, full and productive employment and decent work for all

SDG 9
Build resilient infrastructure, promote inclusive and sustainable industrialization and foster innovation

SDG 10
Reduce inequality within and among countries

SDG 11
Make cities and human settlements inclusive, safe, resilient and sustainable

SDG 12
Ensure sustainable consumption and production patterns

SDG 13
Take urgent action to combat climate change and its impacts

SDG 14
Conserve and sustainably use the oceans, seas and marine resources for sustainable development

SDG 15
Protect, restore and promote sustainable use of terrestrial ecosystems, sustainably manage forests, combat desertification, and halt and reverse land degradation and halt biodiversity loss

SDG 16
Promote peaceful and inclusive societies for sustainable development, provide access to justice for all and build effective, accountable, and inclusive institutions at all levels

SDG 17
Strengthen the means of implementation and revitalize the global partnership for sustainable development

SHAPING
THE FUTURE

The route
to success

**ICE's Sustainability Route Map is helping
to keep the industry on track**

With only 10 years left in which to deliver the United Nations' Sustainable Development Goals (SDGs), attention has focused on the risk of inactivity and the need to deliver actions now.

The World Economic Forum in Davos, Switzerland in January 2020 saw high-profile figures such as HRH Prince Charles and Greta Thunberg challenging global leaders to tackle inertia, with the Prince of Wales launching a Sustainable Markets Initiative. "Sustainable markets generate long-term value through the balance of natural, social, human and financial capital," said the Prince. "Systems-level change within sustainable markets is

> *"We have a unique opportunity at ICE to inspire our members to deliver a more sustainable built environment, and we also have a duty to provide them with the knowledge, skills and expertise in order to do so"*

driven by consumer and investor demand, access to sustainable alternatives and an enhanced partnership between the public, private and philanthropic sectors."

A similar shift is required in the built environment, especially as infrastructure is at the heart of global development and delivering the SDGs. A recent paper for the journal *Nature Sustainability* found that infrastructure either directly or indirectly influences the attainment of all of the SDGs, including 72 per cent of the targets. "We therefore have a unique opportunity at ICE to inspire our members to deliver a more sustainable built environment," says David Balmforth, Chair of the Sustainability Route Map Steering Group, "and we also have a duty to provide them with the knowledge, skills and expertise in order to do so."

Delivering the SDGs is one of the greatest challenges faced globally. The risks of failure are catastrophic, but the possibility of a sustainable, peaceful and prosperous planet is the reward if ICE and other organisations are able to deliver.

MAPPING A SUSTAINABLE FUTURE

ICE's commitment to delivering the SDGs was brought to the foreground in 2018 with the institution's largest ever event; the Global Engineering Congress (GEC). The GEC was inspirational in demonstrating how engineers might better shape a sustainable world. The lessons learnt gave a clear steer on what needs to be done. But perhaps more importantly it established a basis for collaboration, between the professions, governments, non-governmental bodies and global organisations such as the UN, the World Bank and the World Federation of Engineering Organizations (WFEO).

Outputs from the GEC helped ICE to generate its Sustainability Route Map, a programme of activity to transform how engineers

RIGHT
Greta Thunberg speaks at
a climate change protest
prior to Davos 2020

engage with the SDGs. The route map sets out the strategy for delivering its vision in several key areas, which Balmforth describes as: "To harness the capability and capacity of the global engineering community to accelerate the delivery of the Sustainable Development Goals for the benefit of society. To bring about a transformation in the delivery of infrastructure through leadership, advocacy, nurturing collaboration and building knowledge and skills."

ICE is looking to take this forward on three fronts. Firstly, in measuring, monitoring and reporting. "If we do not monitor and measure, we will not know if our interventions are delivering the desired change," says Balmforth. "Reporting is also important because it focuses the mind. Having to report on progress and having sound data in the public domain changes attitudes and priorities."

Secondly, in terms of capacity, capability and education. "Making progress on the SDGs against the future challenges of climate change, carbon reduction and poverty and population growth requires new skills and greater capacity and capability, especially in countries where engineers are less likely to have the support of effective professional bodies locally," says Balmforth. "ICE will work with engineering organisations globally, governments and non-government bodies, charities and trusts to develop the necessary engineering capacity, knowledge and skills."

Thirdly, there is a systems approach. "Infrastructure is heavily networked and interconnected," says Balmforth. "Sustainable development of infrastructure therefore requires a systems approach. Existing systems capability can be extended into new geographies and new contexts."

The infrastructure challenge

Well-considered infrastructure and interconnected thinking are critical when it comes to delivering on the sustainable development goals

In January 2010, a 7.0 magnitude earthquake hit Haiti, causing the death of around 200,000 people and displacing 1 million residents. Six years later, an earthquake of the same magnitude in Japan resulted in more than 40 deaths and displaced around 110,000 people. While neither incident was free of casualties, the degree of impact significantly differed between the two. A key reason for this was the quality of the surrounding infrastructure.

"Infrastructure plays a critical role in society because it can influence development far into the future, both positively and negatively," says Grete Faremo, Under-Secretary-General and UNOPS Executive Director. "This is why it is vital to understand the influence infrastructure systems have on the Sustainable Development Goals."

Infrastructure impacts up to 92 per cent of all Sustainable Development Goal (SDG) targets. Good-quality infrastructure not only enables development, but protects it. Poorly planned, poorly implemented or poorly operated infrastructure can have severe consequences, including destroyed habitats, pollution, overexploitation of resources, and the perpetuation of inequality, vulnerability and fragmented communities – not to mention loss of life and livelihood.

ENSURING A SUSTAINABLE IMPACT

These consequences are intensified by unprecedented stresses – climate change, rapid urbanisation and economic instability – on ageing infrastructure systems. Due to the long lifespan and high cost of infrastructure, poorly planned policies and investments can lock in unsustainable practices for decades. With so much at stake, making the right infrastructure choices has never been more important.

"Due to the long lifespan and high cost of infrastructure, poorly planned policies and investments can lock in unsustainable practices for decades"

"Every piece of infrastructure interacts with its surroundings, which makes it part of a complex system of interrelated elements"

Providing basic services like healthcare, education and justice does not just involve building hospitals, schools and courts. It is also about systematically analysing the problem to find an integrated solution. An example of this approach is the Darfur Urban Water Supply Project, which supplied clean water to more than 240,000 people across western Sudan.

Following years of sustained conflict, a majority of the urban population in Darfur, Sudan lacked access to clean water. The ageing water infrastructure, poor planning and lack of funding could not accommodate the challenges associated with population growth and climate change. The Darfur Urban Water Supply Project – funded by the UK Department for International Development and implemented by UNOPS with the government of Sudan and the Darfur state authorities – began in 2010 to improve access to clean water in four state capitals.

Alongside the rehabilitation of key infrastructure, works included the rehabilitation of 42 deep boreholes, the construction of four major pump stations with storage facilities, 56 km of distribution pipelines and the installation of two chlorination facilities. In addition, staff were trained, new financial-management systems were installed and improved power supply was provided to ensure that the pumps worked well.

BELOW AND OPPOSITE:
The Darfur Urban Water Supply Project delivered an integrated solution to the challenge of providing access to clean water

Ultimately, the project impacted not only SDG 6 (clean water and sanitation), but multiple others, including:

- SDG 1: No poverty
- SDG 2: Zero hunger
- SDG 3: Good health and well-being
- SDG 7: Affordable and clean energy
- SDG 8: Decent work and economic growth
- SDG 10: Reduced inequalities
- SDG 12: Responsible consumption and production
- SDG 13: Climate action

The Darfur project illustrates that sustainable development depends on interconnected thinking. While the project's main output was clean water, a sustainable supply was not achieved solely through the provision of water – it also required cross-sectoral thinking. This involved looking beyond the physical assets and examining the complex network in which they operate.

INTERCONNECTED THINKING

No piece of infrastructure exists in isolation. Perceiving it as such risks weakening communities' resilience to shocks and stresses. For example, a January 2020 UNOPS report, "Infrastructure and Peacebuilding", highlighted correlations between poor infrastructure development and increased vulnerability to conflict in fragile and conflict-affected areas.

Every piece of infrastructure interacts with its surroundings, which makes it part of a complex system of interrelated elements.

These elements include not just the building or bridge, but also the policies, procurement procedures, regulations, financial and enforcement mechanisms, and codes and standards that govern those assets' life cycles. The following elements form an infrastructure system:

- Asset: The physical piece of infrastructure
- Institution: The methods and frameworks governing and regulating the asset and its performance. For example, how was the need for this asset identified? How was it designed and constructed? How is it being operated and maintained?
- Knowledge: The expertise of individuals in the planning, design, implementation, use and maintenance of the asset. For example, who was involved in the asset's design and construction? What are their skills?

Infrastructure systems also interact with other systems in their given context. In this way, various infrastructure systems form a network – a system of systems.

The Darfur project addressed all aspects of the infrastructure system by constructing assets (such as wells, pumps and pipes), supporting the institutional environment by improving financial collection and management systems, and developing the local community's knowledge and skills through training. The project also looked beyond the water sector, improving energy supplies around the boreholes to ensure water service delivery. Altogether, this reduced households' expenditure on water and provided more than 240,000 people with access to a clean water supply.

Infrastructure systems impact far more than individual sectors – they can be the foundation of sustainable development. Only when decisions consider the interdependencies between systems and sectors will infrastructure development lead to a prosperous and sustainable future for all.

ECONOMIC DEVELOPMENT AND INNOVATION

Investing in people

Inclusivity, employment rights and good working conditions play an important part in the ILO's sustainable approach to infrastructure development and environmental works

As the only tripartite UN agency, the ILO brings together governments, employers and workers of its member states to set labour standards, develop policies and devise programmes promoting decent work for all women and men. It promotes rights at work, productive employment, social protection and social dialogue. All are essential for achieving the goals of personal fulfilment, decent living standards, and economic and social integration and development.

Within the ILO, the Employment Intensive Investment Programme (EIIP) supports countries in the design, implementation and evaluation of policies and programmes to address unemployment and underemployment through public investments, typically in infrastructure development. It links infrastructure development and green works with employment creation through the use of local resource-based approaches. These approaches combine and optimise the use of local resources such as labour, skills, materials and technologies, and are implemented in close collaboration with local authorities, contractors and communities.

The construction industry is an important source of employment, accounting for 7.6 per cent of global employment – the equivalent to around 230 million jobs. Construction typically contributes 5–9 per cent to GDP in developing countries and has strong backward and forward linkages to the rest of the economy. Because of this, investing in infrastructure has significant potential to create jobs, particularly in developing countries, where additional infrastructure is most needed.

A large proportion of workers in these countries are often unskilled and in the informal economy, and for many construction

offers an entry point to paid and formal employment. The jobs created are decent as they are informed by good labour practices, observe national labour legislation and pay attention to occupational safety and health. The EIIP provides a model for infrastructure development that is inclusive and can improve the lives of vulnerable people, including women, youth, refugees and internally displaced people and people with disabilities. EIIP is always working with its social partners to ensure that their views are closely reflected in shaping its programmes.

The EIIP works primarily in low and middle-income countries, and with small and medium-sized contractors. Experience shows that the approaches it promotes are particularly suitable for the construction and maintenance of infrastructure, such as rural roads, irrigation schemes and small earth dams, flood protection and river control, social infrastructure, potable water supply and rain water harvesting.

Engagement with small and medium-sized contractors, including community contractors, has various benefits, including the rapid and cost-efficient mobilisation of services, direct engagement of local labour, procurement of local materials and equipment, and long-term working relationships between local contractors and clients. Labour-intensities of the above works, or the percentage of project cost that goes to labour, can be as high as 70 per cent depending on the types of infrastructure, the cost of labour and country characteristics.

Environmental works also have a good potential for employment creation and are by nature labour-intensive. EIIP defines physical

activities that have an overall positive environmental impact as green works. Key sectors for green works include irrigation, flood control, soil and water conservation and forestry. Labour intensities can increase by up to 35 per cent when introducing local resource-based approaches and bioengineering. As such, local resource-based works often constitute a key strategy for socio-economic development and environmental restoration.

Increasing the use of local resource-based approaches around the world can contribute to achieving multiple Sustainable Development Goals (SDGs): by generating employment and increased income (SDG 8); supporting investments in roads, irrigation, dams, markets, schools and health centres (SDG 9); and community, land, ecosystems and environmental assets (SDG13). It can also sometimes contribute to peacebuilding (SDG16). This is even more relevant now, as the attainment of the SDGs will be more challenging due to the deteriorating socio-economic situation in many countries caused by the COVID-19 crisis.

EIIP programmes are, in fact, also often implemented as a response to a crisis or to an economic shock. In various countries, the response to the economic fall-out of the current COVID-19 crisis will likely include stimulus packages including investments in infrastructure. If well designed, such stimulus can improve and maintain infrastructure, help with the recovery of ecosystems, address the climate- change agenda and create employment simultaneously.

INCLUSIVE RURAL DEVELOPMENT
An ILO infrastructure project in Timor-Leste illustrates how the agency's approach works in practice. Seventy per cent of the country's population lives in rural areas and promoting sustainable employment

PREVIOUS PAGES
AND OPPOSITE

Infrastructure policy
and projects play an
important part in the ILO's
promotion of workers'
rights and support for
female employees

and growth in its rural areas is key to the achieving of its development targets. To help, the ILO is creating work for vulnerable men and women, including those with disabilities, in the agro-forestry sector.

Rui Guterres lives in Baguia, in eastern Timor-Leste. He struggled to support his family of three children by a mixture of rice farming and rearing animals, and hoped for a job that would provide a more regular income. However, when, in September 2018, he heard of a road-building project in his village, he thought his chances of being hired were limited.

Rui, 32, was born with a damaged right leg. "People are all different, but being disabled is a whole different ball game," he said. "Still, I went and attended meetings in the village to learn more about the project and, who knows, take a chance."

At the meetings he learned that more than 90km of rural roads in his area were to be rehabilitated and maintained by Timorese contractors, as part of the ILO-led Enhancing Rural Access Agro-Forestry Project (ERA Agro-Forestry). The EU-funded project aimed to promote rural development and social inclusion.

"The workforce for the rehabilitation and maintenance projects is recruited from the communities adjacent to the roads," explained Albert Uriyo, project manager of the ILO project. "The project benefits local communities by boosting local economies through wage transfers, building workers' skills, and providing an opportunity for individuals – in particular the most vulnerable, such as people living with disabilities, to gain experience with more formalised employment."

The ILO project focuses on disability inclusiveness and gender equality. The inclusive approach is part of its Social Safeguards Framework, which includes health and safety-related provisions and equality and child equality provisions. A particular focus on disability inclusiveness and gender equality is incorporated in the agreements signed between the local communities benefitting from the roads and the contractors – around 40 so far – carrying out the work.

Rui landed himself a job as a low-skilled worker with one of the small contractors implementing the project, which had received training from the ILO Project. "I am very happy to be able to have

a regular monthly income and because of it, I can now fulfil my family needs," he said. "My family now lives better than before because their daily needs are being met unlike before."

Another of the project's social targets was to ensure that a third of the employment days (totalling 450,000) benefitted women. Maria Angelina Guterres – not related to Rui – was one such beneficiary. The 27-year-old mother works as a supervisor for one of the scheme's contractors.

Before taking the job, she worried about the challenge it would present because she had no training in labour-based road maintenance. But she was able to take a two-month training programme and learned techniques for business and contract management. "I'm now equipped with the skills I need to carry out my duties as a supervisor," said Maria. "I'm also being recognised for my ability by my peers."

As Albert Uriyo explained, "Women in Timor-Leste continue to be underrepresented in the labour force, particularly in rural areas where formal employment opportunities may be limited. Perceptions about women's capabilities, domestic responsibilities and physical demands of certain types of work may also discourage them from seeking employment outside of the home."

Since June 2017, the project has trained 25 contractor companies in effectively implementing and supervising road construction projects that ease access to markets and remote areas. Coaching and mentoring to ensure the effective transfer of skills are part of the programme.

However, challenges remain for women in "technical" fields and those with higher skill levels. Maria was one of only three female supervisors, out of a total of 46, trained by ERA Agro-Forestry in the municipality of Baucau. "It is an uphill challenge, which requires greater participation by women in science, technology, engineering and mathematics, and efforts to encourage more women to participate in technical fields," said Albert.

"This will develop employment, income support, skills, and access to agro-forestry product markets," he concluded. "The key is to empower local populations in an inclusive and sustainable development process."

The road to long-term economic development

Stable infrastructure is crucial for developing countries to achieve economic growth, as a recent EU-funded project in The Gambia, implemented by UNOPS, goes to show

Sustained and inclusive economic growth is a prerequisite for sustainable development. It contributes to improved livelihoods globally, leading to new and better employment opportunities, and provides greater economic security. Among the least developed and other developing countries, rapid growth can help reduce the wage gap relative to developed countries, diminishing inequalities between the rich and poor.

Infrastructure systems play an important role in increasing national economic growth and productivity. At a fundamental level, infrastructure can help reduce poverty by providing a range of basic services such as electricity, water and waste disposal. Infrastructure networks facilitate the provision of these services, and sustained provision ensures communities have the means to recover swiftly from social or environmental shocks without falling back into poverty.

CONNECTED THINKING

Making economic growth more inclusive means reaching and providing economic opportunity to all, including the most vulnerable. For example, UNOPS is implementing a programme

– funded by the European Union and carried out in collaboration with the National Roads Authority and Ministry of Transport, Works and Infrastructure – to rehabilitate more than 100 km of rural roads in three regions of The Gambia. The road network will link rural villages with major motorways, connecting remote areas to towns and cities across the country.

The project also provides technical assistance and training on labour-based construction methods to improve the economic impact of construction projects and to ensure the country's road network can be effectively managed and maintained. By improving means of travel, it will also help facilitate access to markets and socio-economic facilities for some of the most vulnerable populations in The Gambia.

"UNOPS is excited to participate in this feeder roads and capacity-building programme that will benefit the end users of the feeder roads and those responsible for road management, including routine maintenance," said Nick Hodgson, UNOPS Country Manager, in 2017. "We are also happy to directly contribute to Sustainable Development Goals 1, no poverty, and 8, decent work and economic growth, as a result of increased economic activity along the feeder roads."

"In the long run, the benefits of strong national infrastructure can support the achievement of outcomes across all 17 SDGs"

Meanwhile, research from the World Bank and the IFC has shown that secure and accessible energy sources, transport networks to link producers and consumers, and digital communications technology, support national economic growth and productivity by increasing value added and providing efficiencies at all stages of the value chain.

In rural or remote areas, digital communications and improved transportation networks can provide access to basic services, supply residents with vital goods and information, and link them to facilities. For example, the Overseas Development Institute has shown how transportation can connect workers with suitable jobs, facilitating commuting for skilled and unskilled workers, including women and youth. The transition to a digital economy also creates new opportunities for growth through a range of ICT-based jobs while allowing for teleworking and financial inclusion through access to digital banking and services.

Access to reliable infrastructure that performs these functions – as well as the provision of built economic infrastructure such as

factories, industrial and storage facilities, and markets – is key to attracting new investment and expanding economic activity while generating employment and improving livelihoods. In the long run, the benefits of strong national infrastructure can support the achievement of outcomes across all 17 SDGs.

Infrastructure system design can also incorporate objectives around industrialisation and industry, sustainability, resource-use efficiency and equitable access. In its feeder-roads project, UNOPS is encouraging more women to work in the construction industry – the project employs a team comprising more than 60 per cent women, providing valuable skills that can open up new career opportunities and improve livelihoods.

"In my community, we have always believed road construction to be for men only, so we were surprised when UNOPS offered us the opportunity to work on this project," said Fatou Sanneh, from Alkali Kunda in the North Bank region of the country. "I would like to continue in this work – maybe one day I will become a supervisor or road engineer."

Building collaborative change

An innovative platform is addressing the lack of diversity in the construction sector by aggregating information on people, organisations and opportunities in a clear and accessible way

SDG: 8

What is the response to a fragmented industry that has homes, buildings and infrastructure to build, yet a lack of skilled workers and a real problem with diversity? Keep working in silos, or come together to enable holistic change aggregating supply and demand, diversity, skills shortages and social value?

"The problem we have is that many organisations limit their focus to one trade, profession, activity or sector," says Rebecca Lovelace, founder and "Chief Dot-Joiner" at Building People. "We are taking a holistic approach by creating connections across the built environment, driving traffic to what already exists and aggregating opportunities to make it easier for the industry to recruit more diversely."

It's an approach that Andrea Vihristencu, Researcher of Skills and Training at BRE (the Building Research Establishment), endorses. "Building People is one of the best platforms to use when looking for innovation and change in the sector," says Vihristencu. "Having valuable content in one place is extremely useful and it is curated in an intelligent and organised manner, with the search engine coordinated around diversity of all kinds, from ex-military to ethnic groups."

The UK needs to build. Around 300,000 new homes are needed every year, HS2 and other major projects are going ahead, construction output is growing by 1.3 per cent a year and 168,500 construction jobs are being created between 2019 and 2023. But there are not enough skilled workers and the predominantly white, male, ageing workforce that does exist is not representative of today's society.

THE DIVERSITY CHALLENGE

The percentage of black, Asian and minority ethnic (BAME) individuals in UK construction is between 4 and 7.4 per cent. Construction has the third-worst image among industries as an LGBTQ+ employer and around 14 per cent of the construction workforce is female, with the proportion working as site operatives a very low 2 per cent. The industry increasingly recognises the importance of having a diverse workforce, with social value climbing the procurement agenda, but there is no aggregated way to find diverse individuals and to connect supply to demand.

"We are seeking to deliver the change needed," says Lovelace, "spearheading a collaborative approach to enable industry to come together around people,

BELOW
Building People is coordinating information and opportunities to address the construction industry's skills deficit

skills, social value and diversity." Starting with a grant in 2018, Building People has developed as a social enterprise to create a "network of networks", which brings together diverse and innovative communities to widen the industry talent pool.

There are many fragmented organisations and initiatives that work with diverse communities. But it can be hard to find information and target different groups, so the Building People platform helps by providing the construction sector with a simple means of engaging with the information that is out there. It helps users find support, such as job brokerages, mentoring, networking or access to work experience. And a searchable directory lists almost 1,000 places to go for individuals to find assistance, and for employers to find potential workers. This can be filtered by audience, for instance, women, ex-military, refugees, young people; by trade or profession; by region; or by activities, such as mentoring, learning and work opportunities.

An example of the aggregating service that the platform provides is that if a user searches the internet for UK organisations focused on, for example, "women in construction", only a couple come up on the first page of results. Looking at a key industry website there are eight listed. But, when using Building People, 86 organisations and initiatives are listed clearly in one searchable place.

The platform has a holistic focus to encompass the many silos within the built environment. Its integrating and underpinning approach means that it enriches content without disrupting user journeys.

Over time, Building People will expand its reach to different occupations across industry, adding in new functionality to connect people to more activities, with the ultimate goal of delivering an aggregating and connecting service for all across the built environment. This will help the sector achieve full and diverse employment, and enable those employed in the sector to reflect the full range of diversity in the population.

Post-conflict reconstruction

When rebuilding after a conflict, infrastructure needs to be carefully considered in order to achieve a sustainable recovery

SDGs: 1, 3, 8, 9, 10, 11, 17

It is almost impossible to imagine what war is like for those who have never experienced it. The media show us images of the destruction, of the wounded and the refugee camps, but rarely provide a sense of what it is like to pick up the pieces after the fighting. The world of those affected is changed for ever, both physically and in the hearts and minds of those left behind. Significant rebuilding is needed, but *how* it is rebuilt will determine whether recovery is sustainable. It can be even more important than *what* is built.

The world saw how communities changed in the Bosnian War – from the ethnic mix of the community to where people settled. It changed the scale and location of demand for utilities and services, which rarely reflected the situation prior to the war. This presents two problems. The first is to understand the complex new reality; the second is how to rehabilitate communities.

These issues have been recognised and discussed for decades, and yet reconstruction conferences still talk about the reconstruction of what existed prior to the destruction, for example the 2014 Cairo Summit on the Gaza Strip. Why is this? It would seem that the extraordinary complexity of post-war rehabilitation requirements encourages their simplification. Interestingly, reconstruction, and that of infrastructure in particular, promises to be the most effective catalyst for rehabilitation. To realise this benefit, we need to understand the new physical realities and view infrastructure as a system that enables society rather than as real property assets.

INFORMED DECISION MAKING
Advances in "stand-off recognition" allow us to understand what the physical situation is after the fighting. Stand-off recognition is the process of analysing satellite images of the conflict area to see how land is being used, where the

> *"For the local community to have a sense of ownership, there needs to be tangible benefit"*

concentrations of human habitation are, what condition the infrastructure systems are in, and even what crops are growing where. It is an extension of the growing discipline of Exploratory Spatial Data Analysis. This was very effectively applied in the Gaza Strip in 2018 to understand the true condition of essential services and land use. The same approach was also successfully applied to support economic development in stable countries not recovering from conflict elsewhere in the region.

Using this process, a "common reference" model can be created. The common reference is an evidence-based model that allows all stakeholders to reference the same reality, while interpretations of need may still differ. It enables informed dialogue and a common recognition of what a proposed action would mean, and by extension where the risks may be. Stand-off recognition allows us to build the spatial model. This can then be overlaid with operational models and political and ethnic maps.

This process in the Gaza Strip enabled local stakeholders to build and use the common reference model, derived from standardised data, to agree on what infrastructure is where and what it depends upon. Machine learning (the use of computer algorithms that improve automatically through experience) and the new generation of satellite imaging have already made the common reference a timely and accessible tool. Artificial intelligence promises to incorporate scenario analysis and risk sequencing for reconstruction planning and financing.

The influence of infrastructure on community function is centred on the idea of community ownership. This is not unique to countries and regions recovering from war. When a community has a sense of ownership of its infrastructure, that infrastructure will be adopted, used and adapted, even if this doesn't resemble what the sponsor intended. The point is that the value of infrastructure lies in what it enables and how it is used, not as an asset. This is what distinguishes it from real property. For the local community to have a sense of ownership, there needs to be tangible benefit for the community, at both an individual and a collective level. It means that workers are not brought in from other regions in preference to locals, and that

the infrastructure projects are accessible to as many people from across the community as possible. This is known as "Beneficial Capability".

BENEFICIAL CAPABILITY

There are three parts to Beneficial Capability. Firstly, households must be able to benefit from the infrastructure, which means that the potential earners in the household must be employable and are available to work. Many households are widowed, actually or effectively, by war and face the choice of caring (for the children, sick and infirm) or earning an income. This is the tipping point for post-conflict poverty. Secondly, the earner must be able to get to their place of work. The offer of employment in Sinai across the Rafah border crossing into Egypt for Gazans, for instance, is only practical if getting there and back can be done in less time than the working day. Finally, the project or facility must be configured to make use of local labour and resources. This aspect is known as Intelligent Resourcing and has been around since the days of the Roman Empire.

These criteria were applied to some of the reconstruction projects in southern Iraq during the Coalition Provisional Authority jurisdiction. Community facilities were established so that the young children of households could be cared for, vaccinated and fed, while the earners engaged in the reconstruction work. No one had to prioritise income over the health and welfare of their dependents. The local projects were typically oriented to suit local resourcing, maximising the benefit for the local neighbourhoods. The subsequent insurgency and national governance challenges notwithstanding, these local reconstruction projects directly benefited those most affected by the war.

These concepts all enable infrastructure reconstruction that is informed by the actual situation on the ground; that involves the whole community; that builds stronger communities; that directly enables the next generation; that alleviates post-conflict poverty; and that is inherently sustainable. That has to be a good thing.

Material benefits

A recent road project in Staffordshire used an alternative, local source of aggregate to reduce CO₂ emissions and project costs

SDG: 9

Highway construction projects often make huge demands on natural resources, from cement for concrete, to gravel for pipe bedding, to aggregate for pavement construction. In December 2018, a new junction, designed to reduce congestion and improve road safety, was opened on the A50 in Uttoxeter, Staffordshire. The project looked closely at ways to reduce environmental impact through the sustainable sourcing of key materials.

In order to take traffic up and across the new bridge over the existing A50, the design required almost 400,000 tonnes of natural granular aggregate to fill the embankments and form the slip roads. The side slopes were designed to minimise the footprint of the scheme on the surrounding landscape.

It was also important to reduce vehicle movements close to residential areas and the land-take on a greenfield site. Through an early buildability assessment, alternatives were investigated to reduce the environmental impact of the construction. One solution was a waste product known as pulverised fuel ash (PFA), a by-product from coal-fired power stations that was available only 15 miles from the site. The relatively local supply was closer than the natural sources of aggregate and had a much lower density (by around 0.5 tonnes per cubic metre) making it much more environmentally and economically efficient to transport.

A SMART ALTERNATIVE

It took some extra design work to make the new material fit within the original road design, but there were many benefits to using it. The low-density PFA reduced the weight of the road material (the load) placed on the existing poor ground, reducing the amount of consolidation that the embankment would undergo and therefore the amount of fill required. The loads predicted in the original design also required the installation of cement-bound strengthening columns known as Deep Soil Mixing. This was a time-consuming, risky and expensive operation that became redundant with the introduction of low-density PFA. Although many of the coal-powered power stations in the UK have now been closed or converted to alternative fuel sources, there are still large lagoons and stockpiles of PFA that have accumulated into waste heaps.

The buildability assessment also suggested that rather than dispose of the unsuitable material from the excavations below the bridge and the cuttings for the road, the project could reuse some of this poor-quality ground material. Instead of transporting 52,000 tonnes of waste to landfill, the material was treated with lime, which reduced its moisture content, making it suitable for use in the embankments and side slopes. The use of PFA and retaining of waste material on site saved more than 6,000 vehicle movements into and out of the site, which equates to around 300 tonnes of CO_2 saved.

Another innovation on the site was the use of a glass sand made from crushed waste glass bottles, which was suitable as a replacement separation layer used in the base of the embankments.

In addition to reduced CO_2 emissions from transport, with fewer wagon journeys required to bring the material in and fewer wagon journeys out to the tip as material was retained on site, there was reduced waste, both at the power station and from the site, and increased predictability of supply from a stockpiled material. Moreover, there were massive cost savings, both from the imported fill (a £2.5 million saving) and from the cost of taking material to landfill by modifying with lime (a £400,000 saving).

OPPOSITE
The use of alternative materials resulted in a reduced environmental impact as well as cost savings for the A50 junction project

Data that
works harder

INTROHIVE
Location: Worldwide
www.introhive.com

The construction industry's problems with productivity are well-known. For all the hard work and innovation that take place in the sector, a report in 2019 described low productivity as the industry's "Achilles' heel" – something that makes sustainability even harder to achieve.

Could the software developed by Introhive be the solution to remedy that challenge? Introhive is used by several leading architecture, engineering and construction companies including Clark Nexsen and Gleeds, as well as many of the world's largest brands in the professional services industry.

Introhive was founded in 2012 by Jody Glidden and Stewart Walchli with the objective to bring relationships to the forefront of business management. Throughout their careers, relationships have been at the core of successful business growth. Mapping those relationships was never easy, and even with the key business systems like CRM (customer relationship management), the task still wasn't straightforward or simple. They would turn to CRM systems for insights into key accounts, but were disappointed with the lack of data-driven insights and actionable intelligence that lived within CRM.

"The bottom line is that CRM is a big investment and the hard truth is people simply don't take the time to use it properly, which hurts a business's ability to grow and maintain their competitive edge," says Introhive's CEO, Jody Glidden. Introhive set out to create a platform that leverages artificial intelligence and machine learning to automate mundane tasks for busy professionals and to identify valuable and previously unseen insights about their customers and winnable business.

Introhive's relationship intelligence and CRM automation technology enable employees to save up to 12 working hours a week. Between automatic contact and activity synchronisation between email exchange and CRM to their patented Pre-Meeting Digest (which delivers relevant meeting information directly to email prior to meetings or calls), previously non-value-added task management is eliminated. "The time savings from automation is one of the fastest ways to increase time spent serving current clients and winning new work," says Glidden. "The time saving per employee is a nice added value, but the downstream impact to efficiency gains, increased revenue and higher profitability is what our customers like best."

Glidden explains that the volume and accuracy of each customer's data is truly transforming both their go-to market success and their sustainability, even during times of economic turbulence. "Introhive can go into an organisation's historical email data and analyse who in the business has been engaging with and who, on the basis of that data, has the best relationships," he says. "It can even analyse businesses that have been impacted by layoffs or furloughs. Introhive can retroactively look back at historical contacts and people engaged with the business to ensure that data isn't lost forever. It is very powerful for organisations experiencing high succession rates or high employee turnover."

"At the end of the day, the value drivers of the platform are universal because all organisations are battling poor data quality and limited visibility into relationships," says Glidden. "Executives, engineers, architects, project managers and work winners will have different reasons why this data is important to them. But the universal outcome is better access to business relationships, insights into winning work and employee productivity gains. Those factors are what fuel revenue growth and sustainability.

Smart software delivers hard results

TOPCON POSITIONING
Location: California, USA
www.topconpositioning.com

Crossrail is one of the UK's biggest and most challenging infrastructure projects in decades, due to the extreme complexity of constructing tunnels and stations beneath one of the busiest and most demanding cities on the planet. One aspect that has made the work a little bit easier – and a lot more accurate – is the hardware and software supplied by Topcon Positioning.

Topcon's "automated total stations" were installed on buildings and used to monitor any movement caused by the tunnelling. The stations supplied by Topcon had one unique feature – they were able to automatically scan and detect newly installed stations elsewhere in the area, which meant that the contractors could proceed with their work without having to repeatedly close roads and access locations to update the software. As one Crossrail engineer says: "It saved many man hours that would otherwise be lost to resetting instruments". The extraordinary accuracy of the instruments was another vitally important factor – they are so sensitive that they can register a movement of 0.1 mm to indicate when excavation has started each day.

It is through instruments such as these that Topcon is working to eliminate waste and improve sustainability in the construction industry, as well as in areas such as agriculture. The original company was formed around 87 years ago and remains headquartered in Japan, where it makes instruments that are used in healthcare. The Topcon Positioning division works out of California, close to San Francisco and Silicon Valley. There it develops a wide range of cutting-edge devices that use GPS to allow for extraordinary precision in a variety of disciplines. The company has been working in this field for almost 30 years, but the potential has ballooned in the past decade since the introduction of smartphones and tablets. "Our location equipment is used by the world's leading digital mapping companies,

that's how accurate it is," explains Jackie Ferreira, Director of Communications. "We have more tech on a bulldozer than you find on a commercial airplane."

Topcon's technology is helping to transform the construction industry. It allows foremen, surveyors and architects to instantly see where any piece of moving machinery – a bulldozer, say, or an excavator – is on a construction site. It allows people to then monitor work in progress to the millimetre, providing instant feedback and allowing decisions to be made quickly and without any waste. "With this sort of accuracy and our advanced software, constructors will no longer be erroneously digging, making mistakes or creating potential erosion issues," says Ferreira. "The amount of planning it allows is incredible. If you always know where everything is located and what it is doing then waste and rework is dramatically reduced. Surveyors can compare their plans with what is actually being built and see any differences to the tiniest millimetre and then accommodate them immediately, in real time. Previously this would have all been done manually. There are now far fewer opportunities to introduce human error, which is why the industry is adopting this software to make their work faster and more accurate."

Ferreira offers the specific example of a road that needs resurfacing to carefully decreed specifications. Rather than close the road and survey it manually, a truck can be driven along the road surveying it electronically using a device developed by Topcon. These readings will capture the precise condition of the road to the finest detail. "You can then make the new surface, customising it to the specific needs and demands of the road, only adding what you need to add where you need to add it," says Ferreira. "There's no waste and it will last longer."

> *"With our advanced software, customers will no longer be erroneously digging, making mistakes or creating potential erosion issues"*

Topcon instruments can even be mounted on drones that allow for the microscopic investigation of dangerous structures or hard-to-access areas such as the underneath of bridges, jobs previously done by men and women operating in hazardous circumstances. These new machines will not cost jobs as there will still be the need for people to operate them. "There is still a huge demand for jobs in construction," says Ferreira. "Technology has become even more important now during the Covid-19 crisis. Contractors must adhere to increased health and safety requirements which can slow down projects and reduce the number of workers on the jobsite. With

technology, the project can be completed while adhering to these new restrictions."

The extraordinary levels of accuracy that these instruments afford will help the construction industry to improve sustainability as resources become increasingly scarce and existing infrastructure becomes old and unreliable. Topcon's work therefore touches on several of the UN's sustainability goals, including SDG 8 (decent world and economic growth); SDG 9 (industry, innovation and infrastructure); SDG 11 (sustainable cities and communities); and SDG 12 (responsible consumption and production).

Topcon's instruments have a wide range of applications outside of construction. They are used in agriculture to help monitor every stage of the cultivation process, from planting to harvest. Again, the precision ensures nothing is wasted, something that will be increasing important as the population of the world continues to grow. "We can map the fields to show exactly where a seed should go and what stage each crop is at," says Ferreira.

"This sort of technology allows precision in every area, from the planting to the feeding to the crop picking. It reduces waste and errors, and allows people to make changes instantly when required." Topcon's instruments and software have even been used in relation to biodiversity and climate change. For instance, they were used to survey the topography of the land during a project to reintroduce prairie dogs to a Native American reservation, and they helped monitor Arctic ice flow and the migration patterns of polar bears in the Arctic Circle.

All of these innovations come together in a mission to eliminate waste and create more sustainable growth for the planet's future. "The pressures that are upon us are huge," says Ferreira. "There are the challenges of an increasing population, of decaying infrastructure and also of infrastructure resilience with the changing weather patterns that mean we are experiencing more hurricanes, floods and tsunamis. Automation in construction will allow us to build better, so what we are building can last longer and protect and serve us better. This argument is about achieving sustainability through technology, which uses accuracy and precision to reduce waste and rework. Now you know exactly where you are building, what you are building and how you are building, so you can reduce mistakes and eliminate waste."

Inspiring trust for a more resilient world

BRITISH STANDARDS INSTITUTION
Location: London, UK
www.bsigroup.com

When the organisation that became the British Standards Institution was formed in 1901 by Sir John Wolfe-Barry, a former ICE president, its mission was to create consistency in an increasingly interconnected and interdependent market. BSI's work ensured that products could operate effectively in different parts of the country and then, increasingly, in different parts of the world.

Over time, this evolved to encompass management disciplines – finding and applying best practice from across the globe in a huge range of industries. Consumer confidence was supplied by the BSI Kitemark, which was established in 1903 as assurance that a product had met BSI's stringent requirements. "It's about consistency and making sure that the standards we write in the UK are applicable, while driving best practice as we become more international," says Dr Scott Steedman, Director of Standards at BSI. "There's no point having a standard that falls down at an international border."

Now BSI, like the UN, is learning from its past to understand the future. BSI was created with the understanding that industry needed a mechanism to create codes of practice and technical specifications. This is still the case today. As businesses embrace sustainability and place greater emphasis on transparency, BSI is well placed to find routes towards consistency and efficiency in these areas using the UN's Sustainable Development Goals as a guide.

"The beautiful thing about the SDGs is they are so simple," says Dr Steedman. "You can see their intent and then decide how you meet them. Companies want their processes to meet these goals and they want to know that consumers are taking their work seriously. And we're there to provide the evidence and supply the confidence through independent auditing. Consumers can make informed decisions, while companies are starting to think beyond shareholder value. It's not just about profit, it's the process and the impact on local communities and the climate."

In embracing the UN's goals, BSI places particular emphasis on supporting trade and innovation and is working at the cutting edge of technologies such as the commercialisation of graphene and smart cities. BSI has been at the forefront of developing international standards in BIM (building information modelling), and the organisation's values of collaboration, transparency and innovation will help define standards and best practice in emerging concepts such as circular economy and the Internet of Things.

"The world is evolving, and the UN, national governments and companies are all realising we need to change our thinking," says Dr Steedman. "One great example is in the Netherlands, where you can't build anything unless you pair it with a building that is being demolished – you have to take around 20 per cent of the old building and put it in the new one. There is groundbreaking work going on in Africa where they are much more advanced in the conversion of light into electricity through photovoltaics. Some of the technologies are far ahead of ours and they use the energy in a much smarter way. We're constantly learning from these cases. Every country has different ideas and needs. It's about being humble enough to recognise good ideas wherever they come from."

Working with 84,000 clients across 193 countries, BSI enables organisations to turn best practice into habits of excellence, serving sectors as diverse as aerospace, automotive, built environment, food and healthcare. Through its expertise BSI improves business performance to help clients grow sustainably, manage risk and ultimately be more resilient and trusted.

Investing in infrastructure

GCP INFRASTRUCTURE INVESTMENTS LIMITED
Location: Jersey, Channel Islands
www.graviscapital.com

The UK government estimated in 2018 that more than £600 billion of capital investment is needed in the UK's infrastructure over the next decade. GCP Infrastructure Investments Limited (GCP Infra) – a FTSE 250-listed closed-ended investment trust – will be looking for opportunities that this presents.

The fund focuses on debt secured against long-term infrastructure projects in the UK which have government-backed revenue support. GCP Infra invests in a diverse range of sectors benefiting from such support. "We focus principally on debt, making us different to many of our peers," says Philip Kent, Director of Gravis, the investment adviser to GCP Infra. "Our objective is to target a number of different sectors, and therefore diversify risk across different asset classes." This offers investors high, stable income with low volatility of returns, while contributing to the growth and stability of the UK.

As government priorities have evolved, so has GCP Infra's portfolio. Around 60 per cent of its current projects are in renewable-electricity generation, but many of the public sector support mechanisms in renewables have expired or are no longer available. GCP Infra is therefore looking for opportunities in other areas, such as renewable heat; or in financing a portfolio of social-housing properties; or in a portfolio of private finance initiative projects in the leisure, healthcare and education sectors.

The role of private capital in the UK's future infrastructure projects is still being determined. "The government has historically focused subsidies on the greening of our electricity generation mix," says Kent. "Heat and transport are areas needing significant investment to achieve the UK's carbon-reduction commitments, such as the charging infrastructure for electric vehicles." The government may also need to implement measures to adapt to the effects of climate change, such as winter floods and high temperatures in the summer.

Whatever the future holds, GCP Infra is well-placed to serve investors and the population alike. Its directors recognise the increasing significance of ESG (environmental, social and governance) concerns for investors. "Infrastructure, by its definition, fulfils a social function," says Kent. "This, combined with the significant focus over the last decade on investment in projects that achieve environmental benefits through emissions reductions, means that infrastructure funds like ours have highly attractive ESG characteristics."

HEALTH, WATER AND FOOD

Building a healthy future

A new sourcebook, published by the World Health Organization, is helping civil engineers put public health at the centre of urban planning

The way in which cities are planned and built can define our quality of life. It affects not only the quality of our living spaces and transport, but also the air that we breathe, the water we drink, and our access to nutritious food, education, healthcare services and employment. Over the years, valuable lessons have been learned about urban and territorial planning, which has developed into a multisectoral discipline. It is now commonplace to consider environmental, social, health and well-being as key determinants when planning cities.

One of the main challenges today is to ensure that urban and regional leaders have the knowledge and guidance to integrate health and well-being into their planning processes. It's why the World Health Organization (WHO), as part of a close and longstanding collaboration with UN-Habitat, has published *Integrating Health In Urban And Territorial Planning*, a sourcebook for urban planners, city managers, health professionals and all those interested in the basis for our collective well-being. The book is designed as a tool to assist national governments, local authorities, planning professionals, civil society organizations and health professionals, by helping to improve planning frameworks and practice through the incorporation of health considerations, at all levels of governance and across the spatial-planning continuum.

The sourcebook also shows how an integrated approach to health can influence decisions on sectors such as housing, transport, energy, and water and sanitation. More importantly, it considers how they are all linked to the 2030 Agenda for Sustainable Development. Health features prominently in the inter-linkages between and among the Sustainable Development Goals, including SDG 11, on sustainable cities and communities, cutting across almost all others and across traditional policy and disciplinary silos. The sourcebook articulates how public health professionals are crucial to good urban and territorial planning. They have a valuable and unique set of skills to bring to the table and can help ensure that routine urban and territorial planning activities, such as economic development or transport planning, are focused on delivering population health and well-being.

THE URBAN PLAN
The skills of public health professionals are particularly relevant as the COVID-19 pandemic continues to highlight the importance of safe distancing in cities. Many cities face health threats linked to urban and territorial planning. Infectious diseases thrive in overcrowded cities, or where there is inadequate access to clean water, sanitation and hygiene facilities; living in unhealthy

BELOW AND OPPOSITE
The WHO's sourcebook aims to provide urban-planning guidance with health and well-being at its core

KEEP THIS FAR APART

environments killed 12.6 million people in 2012 and air pollution killed 7 million people in 2016. However, only one in ten cities worldwide meet standards for healthy air.

"If the purpose of urban planning is not for human health, then what is it for?" says Dr Maria Neira, WHO Director, Department of Environment, Climate Change and Health. "Ideally, cities are planned for adequate standards of living and working, sustained economic growth, social development, environmental sustainability, better connectivity. But the 'why' at the core of all these things comes down to physical and mental health and well-being."

It's a view echoed by Dr Nathalie Roebbel, WHO Unit Head, Air Quality and Health. "Investments in health-based urban and territorial planning secure long-term health and well-being legacies for a growing proportion of humans," she says.

Over half the world's population now lives in cities, and by 2050 that is expected to rise to a full 70 per cent of the human population.

However, 75 per cent of the infrastructure that will be in place by then has not yet been built. This presents an opportunity to build transformative urban areas, especially as the world begins to build back with a greater consciousness of the links between space and health.

One essential consideration is equity as there are substantial differences in health opportunities and outcomes within and across urban areas. The sourcebook is based on the premise that public health and urban planning both aim for fair and equitable outcomes and access to essential services.

"Urban and territorial planning provides a framework to align and transform our built and natural environments," says Laura Petrella, UN-Habitat Chief of Planning, Finance and Economy. "Putting human and environmental health back into the core of the urban and territorial planning process and principles will enable the full potential of our cities and territories to deliver healthier and resilient environments."

Breaking the hunger cycle

The UN World Food Programme works closely with engineers to help deliver resilience and food security

Delivering aid in the midst of a crisis or building food security for vulnerable communities requires reliable, sustainable infrastructure. The lack or poor quality of roads, bridges, ports or airfields hampers the efforts of humanitarian agencies during emergencies. It also hinders development, leaving communities isolated and making access to food difficult and dangerous.

Thanks to a diversified, highly qualified technical capacity, specialised tools and a "can do" mentality, the World Food Programme (WFP) develops tailored, cost-effective and timely solutions to all challenges. Whether it is rehabilitating roads in South Sudan to improve food security and access; preserving food quality in Afghanistan through the design and construction of a Strategic Grain Reserve with silos in five locations across the country; building specialised infrastructure for school feeding in Malawi; reducing CO2 emissions through the Energy Efficiency Programme (EEP); or teaming up with other UN agencies to build and maintain infrastructure in the Cox's Bazar refugee camps in Bangladesh, WFP engineers provide services in remote locations under extremely challenging circumstances.

This function benefits not only WFP operations, but also the wider humanitarian community. This was seen during the 2015 Ebola outbreak, when WFP shared its engineering expertise with other humanitarian partners to build structures that hadn't been built before – from Ebola treatment units to vehicle decontamination centres. In Kutupalong, Bangladesh, one of the world's largest refugee camps, WFP Engineering has built storage facilities to support logistics, humanitarian camps for UN workers and modular bridges to secure access to food and non-food items. In Yemen, WFP Engineering has rehabilitated and constructed essential infrastructure for the humanitarian community to scale up assistance.

Engineering work can significantly contribute to cutting the cost of delivering assistance and has exceptionally high returns on investment. In South Sudan, WFP rehabilitated over 382km of roads from Juba to Rumbek, an area that becomes impassable during rainy seasons. WFP managed to transport 15,000 metric tonnes of food along this road during the rainy season and prepositioned around 132,000 tonnes of food in 2018. If the road had not been open and this food had been transported by air, an additional cost of $37.5 million would have been incurred.

WFP's engineering work bridges the divide between humanitarian and development activities. Whether stemming from an emergency or the request of a government for technical assistance, an investment in infrastructure is an investment in a country's long-term food security.

BUILDING RESILIENCE

Assisting almost 100 million people in around 83 countries each year, WFP is one of the world's leading humanitarian organisations, saving lives and changing lives, delivering food assistance in emergencies and working with communities to improve nutrition and build resilience.

One in nine people worldwide still do not have enough to eat, and the international community has committed to end hunger, achieve food security and improve nutrition by 2030. Food and food-related assistance lie at the heart of the struggle to break the cycle of hunger and poverty. On any given day, WFP has 5,600 trucks, 30 ships and nearly 100 planes on the move, delivering food and other assistance to those in most need. Every year, it distributes more than 15 billion rations at an estimated average cost per ration of $0.31. These numbers lie at the roots of WFP's unparalleled reputation as an emergency responder, one that gets the job done quickly at scale in the most difficult environments.

WFP's efforts focus on emergency assistance, relief and rehabilitation, development aid and special operations. Two-thirds of its work is in conflict-affected countries where people are three times more likely to be undernourished than those living in countries without conflict.

In emergencies, WFP is often first on the scene, providing food assistance to the victims of war, civil conflict, drought, floods, earthquakes, hurricanes, crop failures and natural disasters. When the emergency subsides, WFP helps communities rebuild shattered lives and livelihoods. It also works to strengthen the resilience of people and communities affected by protracted crises by applying a development lens in its humanitarian response.

Climate change, environmental degradation, water scarcity, disease, rapid population growth, unplanned urbanisation: in today's world, heightened risk and fragility are threatening to reverse major development gains. Shocks and stressors such as conflict, natural hazards and political instability can have a devastating impact. Children who are malnourished in their first thousand days of life

may suffer cognitive and physical impairment. In times of war or disaster, schools are the first to close. Historically, humanitarian interventions have saved countless lives and restored the livelihoods of millions. But they have rarely tackled underlying vulnerabilities.

It is true that development programmes are hard to implement in fragile or deeply impoverished contexts, prone to recurrent crises. But evidence suggests that, by embedding resilience in their interventions, development actors can lessen the effects of shocks and stressors, and thus more durably relieve human suffering. For its part, by adopting a resilience perspective, the humanitarian community can ensure that people rebuild better after disasters. Resilience measures, in fact, are cost-effective on two counts: they reduce the need to spend on cyclical crisis response, while helping overcome a legacy of development gaps.

Thanks to half a century of experience, WFP has acquired a comparative advantage in building resilience for food security and nutrition. It has invested in early-warning and preparedness systems – including supply-chain management, logistics and emergency communications – that allow governments to prevent crises or respond quickly when they happen. It is helping to develop national capacities to manage disaster risk through finance and risk-transfer tools, such as weather risk insurance. Its expertise includes vulnerability analysis and mapping, as well as support to social protection systems. In several of its operations, it has developed productive safety nets through community-based asset-creation programmes. Through the Food Assistance for Assets programme, beneficiaries receive food assistance while building or rehabilitating assets such as forests, water ponds, irrigation systems, and feeder roads that will strengthen their resilience and food security in the long term.

This growing body of experience has informed the WFP's understanding; it is now helping shift the organisation's practice. These days, wherever possible, a "resilience lens" is applied at the stage of programme design, and subsequently at all stages of the programme cycle. WFP has learned that no two settings are alike, and that long-term collaboration is crucial. In each distinct context, it must determine how actions can be best layered, integrated, and sequenced with the strategies of national governments and the programmes of partners. WFP's current transition to Country Strategic Plans, whereby national needs and priorities are jointly assessed and agreed with governments and local stakeholders, must be seen in this light: they provide a long-term planning framework that allows WFP to put resilience-building at the heart of its programmes.

PREVIOUS PAGES
AND OPPOSITE
Road-building in South
Sudan (previous pages)
and monsoon preparations
in Bangladesh (opposite,
top and bottom) illustrate
how WFP's engineering
expertise can help combat
illness and hunger

Powering a healthy network

A UNOPS-implemented initiative to strengthen the energy infrastructure in rural Sierra Leone is helping to prepare the region for future health emergencies

In 2014, Sierra Leone faced a medical emergency with the outbreak of Ebola in West Africa. Confronted with more than 14,000 cases in the country between 2014 and 2016, nearly 4,000 people lost their lives. Lack of power in small towns and villages worsened the conditions in which medical professionals worked to fight the outbreak.

For years, community health centres across rural Sierra Leone have been forced to operate without reliable sources of power – an all too common reality for hundreds of thousands of people, putting patients at risk and making the jobs of healthcare workers all the more challenging.

Access to affordable, reliable and sustainable energy is crucial to achieving the Sustainable Development Goals, including SDG 3 (health and well-being) and SDG 6 (clean water and sanitation). Healthcare services provided in hospitals and clinics require reliable access to energy as well as water and sanitation, solid-waste management, transport and digital communications infrastructure.

In keeping with this, the government of Sierra Leone launched the "President's Recovery Priorities" as part of its Ebola recovery efforts. The multi-stakeholder programme includes initiatives to boost energy generation and increase access across the country.

One of the projects developed to support this was the Rural Renewable Energy Project, funded by the UK's Department for International Development and implemented by UNOPS. The project

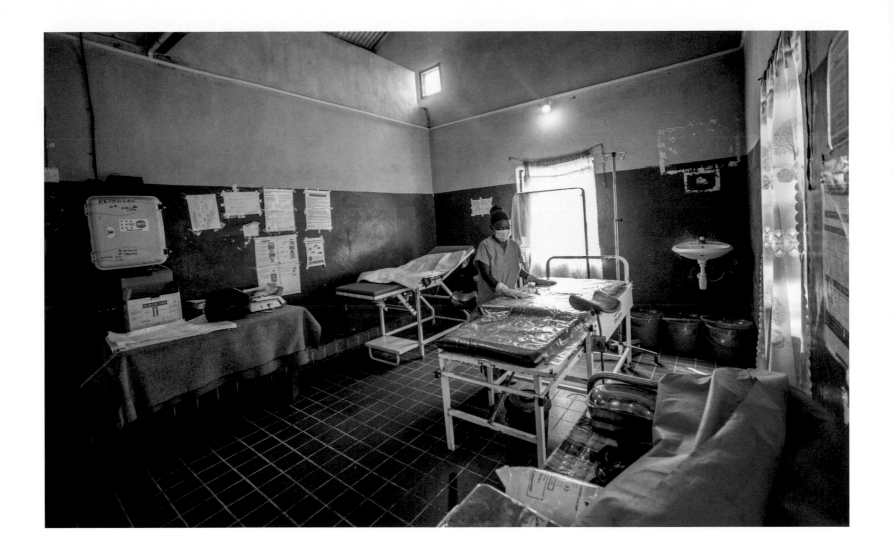

OPPOSITE AND ABOVE
Improved energy provision
in health centres across
Sierra Leone is benefiting
the health and livelihoods
of the wider community

aims to strengthen energy infrastructure in rural areas, improving access to essential services for over 300,000 residents and better preparing them for future health crises.

AN INTEGRATED APPROACH

Quality health infrastructure is delivered through more than just hospitals. Infrastructure networks provide the conditions needed to not only use advanced medical equipment and technology, but also to provide access to quality services.

For example, improved drinking water, sanitation and hygiene play a vital role in maintaining good health. Sanitation services – delivered through clean-water provision and waste disposal – can limit the spread of diseases and decrease the incidence of maternal, neonatal and child mortality by reducing the spread of water-borne pathogens.

Transport networks are crucial where access to facilities is not possible – not only in providing health services, but also in prevention through education and public awareness, and in delivering water by trucks in regions where this may be the more feasible option. Similarly, just as energy security can increase access to healthcare, it can also form a component of water supply, such as through the operation of pumps.

In an increasingly connected world, information and communication technologies can also provide access to health services by facilitating the sharing of knowledge and supporting self-management of medical conditions. Digital communications infrastructure can facilitate recruitment and training of the health workforce and the timely dissemination of information.

The Rural Renewable Energy Project uses an integrated approach to not only address the critical need for electricity in health facilities, but also enhance energy security, support business start-ups, reduce local pollution, and improve the livelihoods and living conditions of local communities – with special consideration for vulnerable groups, including women and young people.

The first phase of the project was completed in July 2017 and involved installing solar power with storage at 54 community health centres in 12 districts. Margaret Albert, a midwife at one of these health centres, says reliable energy has contributed to a more efficient working environment, ultimately enabling more people to receive healthcare. "A lot of deliveries happen at night so the electricity makes the work easier to see when any sorting of medical supplies is being done and to avoid any mistakes or unnecessary discomfort to the mother," she says. "Due to the constant electricity, a good number of supplies are kept in proper condition and this also makes outreach to neighbouring communities to offer inoculations and related services easier."

Through its integrated approach, the project is providing families, businesses and communities with the opportunity to grow, develop and improve their quality of life in a sustainable way.

Sowing the seeds of a sustainable food system

UNOPS is providing its infrastructure expertise in Niger to improve the country's agricultural output and management

More than 40 per cent of the population in Niger lives below the global poverty line of less than $1.25 a day. Despite improvements in recent years, the country's economic growth has been impacted over the past 50 years by climate shocks, political instability, poor governance and regional insecurity.

Extreme and variable climatic conditions exacerbate this situation. Droughts and floods undermine efforts to build resilience and economic security. More than 90 per cent of the population's income depends on the agricultural sector, which is subject to the irregular three-month rainy season and fluctuations in commodity prices that hinder national economic development.

As a part of an effort to reduce poverty, increase food security and improve economic growth in Niger, UNOPS is working with Niger's government to enhance agricultural productivity through infrastructure development. The second Sustainable Development Goal (SDG 2) seeks sustainable solutions to end hunger in all its forms by 2030 and to achieve food security, aiming to ensure that everyone everywhere has enough good quality food to lead a healthy life. Ensuring the secure provision of food and nutrition for all will require the widespread promotion of sustainable agriculture, which can be supported by infrastructure contributions in all sectors.

BETTER PRODUCTIVITY THROUGH INFRASTRUCTURE
Ending hunger through sustainable food systems should account for each step in the chain of food production and consumption: growing, harvesting, processing, storage, distribution, preparation, consumption and disposal. Lack of access to water can inhibit agricultural productivity and food security, which is a key constraint to economic growth in Niger. Agricultural productivity in the country has been impacted by outdated irrigation systems and a lack of reliable water availability. Previously productive areas have deteriorated because of a lack of sustainable management of natural resources.

Improved water supply and irrigation techniques are key to strengthening agricultural resilience to drought and maintaining food supply. Moreover, climate-resilient agricultural systems are essential to ensure adaptability. Agricultural-land-use management can also improve productivity and resilience to flooding.

Modern energy infrastructure plays an important part in streamlining the production process: electrification of food processing can increase productivity through automation, while refrigeration can help reduce food waste and enhance food quality and freshness. Constructing warehouses and other storage facilities also helps manage the food production process and reduce waste throughout the supply chain.

In addition, digital communication provides farmers with the knowledge necessary to improve agricultural techniques. It can play a larger role in food distribution as producers gain access to market information domestically and internationally.

Finally, transportation also plays a key role in agricultural productivity. More efficient transportation linking producers and consumers will increase the sustainability of the food system and the variety of food options available. Adequate links and transport facilities can also ease access to crucial resources.

For instance, agricultural productivity in Niger has been impacted by the lack of access to inputs such as seeds and fertilizer. Lack of physical access and institutional barriers to trade create high transaction costs and lower sales volume.

To address this and other challenges, the Millennium Challenge Corporation (MCC) and the government of Niger developed the MCC compact programme with funding from the US. The initiative supports sustainable water management and agriculture activities that will benefit around 3.9 million people.

The compact programme aims to increase rural incomes through enhanced agricultural productivity by promoting the sustainable use of natural resources while improving access to markets and inputs, such as seeds and fertilizers, irrigation systems and access to water. As a consultant, UNOPS is providing programme and project management support services, including more than 2,500 days of technical assistance in infrastructure and project management in 2017.

By upgrading infrastructure across multiple sectors, the compact programme is tackling agriculture challenges in Niger and addressing key constraints to economic growth in the process. Across four regions, the programme is supporting the growth of agricultural enterprises and increasing market sales of targeted commodities with the aim of increasing the incomes of small farms and pastoralists.

It also aims to improve crop and livestock productivity, increase yields through modernised agriculture and encourage sustainable management of natural resources. Together, these actions will support long-lasting growth and development for Niger's agriculture sector and economy, ultimately providing sustainable solutions to advance the achievement of SDG 2 targets and end hunger.

Remote control

The MANTIS pump-monitoring system provides a simple and effective means by which to check the performance of water pumps

SDGs: 1, 6

"There is currently only limited information on how many of the millions of handpumps installed across the world are actually still doing their job, and visiting each one regularly is really impractical given their hard-to-reach locations," says Dr Andrew Swan, a Reader at Leeds Beckett University's School of Built Environment, Engineering and Computing. "Remote monitoring could really help towards putting more effective and efficient maintenance in place."

To this end, Leeds Beckett University and Environmental Monitoring Solution Ltd (EMS) have jointly developed MANTIS (Monitoring and Analytics to Improve Service) – a water pump monitoring technology aimed at increasing the visibility of pump operation and performance.

Many communities across the globe rely on hand-operated pumps for their water supply, with an estimated 184 million people in rural sub-Saharan Africa and more than 400 million people in India dependant on handpumps.

However, it is widely reported that many handpumps are broken or deliver a poor level of service, causing hardship, disease and even death in the communities that are reliant on them. These problems are well documented. For example, one previous study (Myths of the Rural Water Supply Sector, RWSN 2010) reported that 20–65 per cent of handpumps installed in a range of African countries were broken or out of use, while another recent report (by Marshall and Kaminsky for the Journal of Water, Sanitation and Hygiene for Development, 2016) claimed that 30–40 per cent of rural water systems in developing regions were failing prematurely.

The reasons for failure are complex, but can include a lack of funds for repairs, limited access to spare parts and/or technical skills, inappropriate project implementation and/or choice of technology, and limited post-construction monitoring and support. Such problems can even lead to new handpumps being abandoned, providing only intermittent, poor-quality service, or being seasonally dependent.

It has been argued that a shift in emphasis from infrastructure provision to service delivery and maintenance provision would be beneficial. However, to be successful, this change needs measurable indicators of efficiency, which is where the MANTIS pump-monitoring system comes in. The system was successfully field trialled in Gambia and Sierra Leone in 2017, with trials pending in 2020 near Kolkata in northern India.

SUCCESSFUL TRIALS

"The simple approach employed by the MANTIS system reduces power requirements," says Professor Pete Skipworth, Managing Director of EMS, "which enables the system to function for longer – we estimate undisturbed for five years or more – and also reduces its production costs. Our field trials in sub-Saharan Africa demonstrated the efficacy of the system, enabling us to fine tune the technology before it is deployed in India."

The system detects whether the pump is in regular use and reports patterns of usage. Its simplicity enables it to have an exceptionally long lifespan compared to other remote monitoring systems that are currently under development. The technology has been designed to function for many years without the need for maintenance or re-powering, sending back daily data on hand-pump usage via an easily accessible web interface, which can be monitored to ensure repairs are made swiftly.

"Remote monitoring has a very practical purpose," explains Dr Swan, "to spot broken handpumps quickly so they can be repaired and people can continue to get easy access to water, even in very isolated areas. But the ease with which the data can be accessed means that this technology can also improve transparency and accountability as to how development funds are spent and how effective they are. Funders – whether countries, NGOs or individuals – can also see whether the pumps they've paid for are working."

ABOVE
MANTIS's remote monitoring
system could help detect faults
in water pumps in developing
regions around the world

ABOVE

Without a robust plan of action, London is set to experience more severe water shortages in the future

Capital ideas for water security

Engineering has a vital role to play in protecting London from the threat of drought and maintaining the city's water security

SDGs: 6, 11

In early 2018, Cape Town was approaching "Day Zero" — the day when the citizens' taps would run dry following a three-year drought. As it happens, Day Zero was avoided thanks to water-use restrictions, cutting water to agriculture and pumping in extra water. However, unless decisive actions are taken, it is only a matter of time before a new water crisis comes. This is true for Cape Town, and for many other cities around the world, including London.

To prevent London from becoming the next Cape Town, water engineers develop and apply mathematical models to identify potential risks to water supplies and solutions to mitigate them. These models are particularly powerful because of their capacity to test and explore shocks and scenarios that have never happened, by subjecting a simulator of a water supply system to those conditions. In the same way that a flight simulator tells a pilot how a plane reacts to turbulence and the manoeuvres of the pilot, the water simulator tells water engineers how water supply security changes under different shocks, such as prolonged drought under climate change.

"Our water simulators show that if no action is taken, London is indeed set to experience more frequent and severe water shortages in the future," explains Edoardo Borgomeo, a water scientist at Oxford University's Environmental Change Institute who has been developing these water simulators. "This is mainly down to population growth, but climate change complicates things further as it will mean more frequent and intense droughts."

Through these simulators, water engineers can also help identify solutions to prevent London from running out of water. Water simulators show that aggressive demand management to reduce consumption and losses in the distribution system (called leakage) is a priority to be implemented immediately. But reducing leaks from London's old water pipes is not an easy task and will not be enough on its own.

Instead, engineers have been thinking about innovative ways to augment supplies. These include building new reservoirs, transferring water from other parts of the country or recycling wastewater. How should London choose between these different alternatives? The city needs something that's not too expensive, that keeps residents happy with the price, taste and appearance of the water, while also reducing the risk of the taps running dry. Finally, water supplies need to be shared with ecosystems, so any solution needs to take into account the needs of the environment.

Deciding which action to pursue is particularly challenging, because the future will be significantly different from anything imagined when water supply systems were first built. Population growth and climate change mean that water engineers can no longer assume that past observations of water availability and demand are representative for the future.

INNOVATIVE THINKING

To address this challenge, engineers focus on designs and strategies that combine robustness — through measures that work well under many possible climate futures — as well as flexibility to ensure we retain the capacity to adjust course and overcome hazards that might materialise. In the case of London, this means planning for an innovative wastewater reuse facility in north London while also bringing in additional water via the Oxford Canal.

And this approach has been used to help plan water resources not just in London, but across England. A nation-scale water simulator helped the National Infrastructure Commission understand and quantify the costs and effectiveness of strategies for coping in the most extreme drought conditions.

The project's Main Trunk Sewer is a gravity-based system that is delivering multiple sustainable gains

A modern sewer for a modern city

The Doha South Sewage Infrastructure Project employs innovative engineering to deliver a sustainable sewerage system

SDGs: 6, 9, 11, 12, 15

The Public Works Authority of Qatar, known as Ashghal, has implemented the Doha South Sewage Infrastructure Project, a world-class solution to upgrade and expand the city's sewerage infrastructure. The project is based in Doha's oldest area, the south catchment, and has been designed to accommodate a projected population growth of an additional 1 million people. From the start it has addressed the serious environmental and health issues that have resulted from undersized and aged sewerage infrastructure.

A major element of the project is the Main Trunk Sewer (MTS), a gravity-based sewerage system that conveys sewage to the Doha South Sewage Treatment Plant. It is designed to assure self-cleansing during low flows without solids accumulating, which would become a maintenance issue. Furthermore, the MTS is large enough to meet projected flows for its 100-year design life without the need for pumps, valves or any other mechanical or electrical components within the sewer or associated structures – again saving on maintenance. MTS was constructed using tunnel-boring machines, which required a limited number of shafts, ultimately reducing adverse impacts to the environment.

The Qatar National Vision 2030 aims to transform Qatar into an advanced country capable of sustaining its own development and providing a high standard of living for its people. Ashghal's Doha South Sewage Infrastructure Project contributes to achieving the Qatari National Vision 2030, by addressing the need for world-class and state-of-the-art infrastructure from the fast-growing population of Qatar.

"This is the first project to achieve CEEQUAL certification in the Middle East," says Maher Al Ajam, Programme Director for Jacobs, the engineering group that has been working on the project. "CEEQUAL is the UK-based sustainability-rating scheme that has helped the project to set a new benchmark for other projects in the region in terms of sustainability, health and safety, teamwork, collaboration and innovative solutions."

The Programme Management Consultant carefully specified materials in the structures to be durable for the maintenance-free 100-year design life in the aggressive sewer environment and Doha's hypersaline ground conditions. The whole project team understood the innovative nature of the project for the region in terms of "sustainability management on mega-projects".

In so doing, the team actively participated in sharing best practices and lessons learned with wider members of the civil engineering sector.

MINIMISING LEAKAGE

The design of the MTS protects groundwater from any potential contamination, with three layers of redundancy to prevent leakage, including grouting on the outside of the tunnel, the tunnel itself and a high-density membrane inside the tunnel made from polyethylene (HDPE, a type of plastic). The Environmental Investigation Agency concluded that, based on these design features, the risk of potential leakage was of "no significance". In construction, diaphragm walls, vertical shaft-sinking machines and grouting in the shaft construction significantly decreased the amount of dewatering required for the project, minimising groundwater depletion and aiding efficiency.

By design, the MTS eliminates the use of water during operation. The MTS replaces the existing sewage pumping stations with a gravity-based sewage system, moving sewerage even at low flows without the accumulation of solids and is simultaneously self-cleaning. To minimise the project's impact on limited local water resources, strict requirements on water use for construction activities were put in place, and water was efficiently used. As per the Dewatering Management Plan and the Water Quality Control Plan, water removed from sewerage sludge (called dewatered water) was treated and reused to minimise use of potable water.

To achieve the required maintenance-free design life of 100 years, the project adopted an innovative lining system for big tunnel sewers by combining the segmental lining and Corrosion Protection Lining. This achieved a significant concrete lining thickness reduction of 200 mm, along with an associated reduction in the excavation diameter of 445 mm, which led to a 20 per cent decrease in spoil to handle. The tunnel segments and access shaft risers were all prefabricated and locally manufactured near the project site. Prefabrication provided precision, and hence efficiency, in terms of material procurement and waste reduction.

This strategic project changed the entire approach to foul sewer drainage in the Doha South catchment. It transformed it from a pumped sewer network to a gravity sewer network, with numerous sustainability gains.

A change of course in urban development

The Grey to Green project is helping to flood-proof Sheffield's post-industrial city centre through the use of sustainable drainage systems

SDGs: 6, 11

The "Grey to Green" strategy is a radical proposal from Sheffield City Council to transform redundant carriageways in the city centre into a network of sustainable drainage and rain gardens. Phase 1 was competed in the spring of 2016 and the completion of Phase 2 is expected in June 2020. The project has improved the city's resilience to climate change, enhanced the public realm and increased connectivity in the city centre. It is now attracting investment in new and existing jobs.

The Riverside Business District (Phase 1) and Castlegate (Phase 2) areas of the city centre suffered catastrophic floods in 2007. The completion of the Inner Relief Road in 2008, which used to come through these areas, offered an opportunity to address these issues in a new way. The overall project will transform 1.2 km of redundant roads into attractive linear public spaces, improving the links between the Riverside Business District, Castlegate/Victoria Quays and the rest of the city centre. It includes innovative perennial meadows, an interlinked Sustainable Urban Drainage System (SUDS), rain gardens, eye-catching public artworks, encouraging sustainable travel, and high-quality paved footways and street furniture. The scheme has been designed to improve the environment, making it easier to walk and cycle. Phase 1 and 2 cover over 1 km of linear SUDS and planting – possibly the biggest retrofit SUDS project in an industrial city in the UK.

The project's sustainability performance was verified by achieving a "very good" construction rating by CEEQUAL – the evidence-based sustainability environmental award scheme – for Phase 1. It won two awards at the CEEQUAL outstanding achievement awards for the Landscape and Water Environment categories.

Environmental improvements through good quality planting are recognised as adding economic value to city centres. Local sustainability projects have concluded that quality public realm and open spaces, along with good sustainable design, improves local businesses by providing higher returns on investment. Landscape quality also helps to create the right business image. Once the decision had been taken to construct the significant SUDS, and make permeable a large area that had been impermeable, it was important to think about the planting scheme for both the SUDS and other soft landscaped areas created by the Grey to Green scheme. The easiest solution would have been to grass over the SUDS. However, the council had the vision to achieve an urban meadow in the middle of the city centre through low-input perennial planting.

REDUCED RUN-OFF

The Grey to Green project has involved removing considerable areas of highway surfaces within a highly impermeable city centre location. This not only reduces the surface area for the run-off water when it rains but also provides the space to manage the remaining areas more robustly. A newly created series of ditches called swale cells provide environments to capture, clean, infiltrate, move and store water.

By reconfiguring the surfaces of urban areas, civil engineers can improve water management, gaining both flood risk and water quality benefits. There are local benefits in the immediate environment but also potentially cumulative benefits for the wider catchment. We need to improve future resilience whenever investments are made in the urban fabric, as has been identified in various studies. These environmental improvements allow climate proofing of drainage networks by building new surface capacity to accommodate future predicted increases in intensity of rainfall. Green infrastructure can help intercept flow paths of water from rainfall, which is particularly helpful within the steep topography of Sheffield. This retaining of surface run-off water also promotes groundwater recharge – returning this part of the catchment to a more natural hydrology and in so doing retaining moisture within the city centre.

The Grey to Green project has applied the principle of using water in the landscape, which has proven a successful element within city centre regeneration. Green infrastructure provides the most robust treatment mechanism for highway run-off water in capturing and breaking down pollutants. As a result, this project can demonstrate clean water becoming an asset to a wider landscape through its movement.

ABOVE
The Grey to Green strategy
has benefitted Sheffield
in terms of environment,
resilience and connectivity

HEALTH, WATER AND FOOD   85

Using data to deliver better dams

Environmental flow assessments can ensure that dams and reservoirs have a sustainable future that doesn't adversely affect the local environment

SDG: 6

Dams and reservoirs can have hugely positive effects on their local environment. They can help communities by providing flood control, irrigation, renewable energy and easy access to potable water. But dams can also modify the natural flow of a river, which may disrupt the local ecosystem. Regardless of location, scope or scale, the simple existence of a dam indicates a need to study the possible adverse effects on the environment. Once determined, it is possible to then provide potential solutions. Also, the ever-increasing risks from climate variability make it more important than ever to have robust reservoirs that secure a clean water supply.

One method to minimise the adverse environmental effects is called an environmental flow assessment, leading to applying an environmental flow regime on a dam. Simply put, the river impacted by a dam or reservoir is seen as an additional "water user" and steps are taken to ensure that that river continues to flow throughout the year, as much as possible. This ensures that it never runs dry, thus protecting the local wildlife habitat.

This concept does not simply maintain the status quo; it can include using advanced-flow regime methods, such as artificial flooding of downstream areas, to promote or speed up natural events such as fish migration. An environmental flow must be developed in order to meet the recommended criteria, while being feasible for an operator to maintain, and not compromising a dam's benefits to the community.

Picture an all-encompassing, or "holistic", approach to environmental river flows that is also concise. The proposed assessment is split into three modules: hydrological, hydraulic and ecological. The first analyses river inflows, seasonal variations, as well as extreme values, meaning long droughts and heavy floods. The hydraulic module locates key points in the river's wetted perimeter, linking specific discharge values to fish food production, and the final module defines several fish quality classes, linking each to a typical flow value and river cross-section. The end result is an environmental flow policy that proposes a dynamic river flow regime.

This includes an original streamflow replica, incorporating artificial flooding by releasing extreme daily flows, and at all other times generally following the river's pre-existing natural regime. The aim is to mirror the conditions that existed in the environment before dams were introduced, while also being a versatile enough scheme for operators. This, combined with the fish classification concept, produces a method that can yield results with a noticeably small amount of inputs. In some regions, the data currently available is severely lacking, and even streamflow measurements required for the most basic approach can be untrustworthy. As data of 10–20 years is required to conduct a valid study, there is an imminent need to start collecting more river data.

GOING WITH THE FLOW

Environmental flows from dams are the ideal way to minimise their negative impacts on local ecosystems. Developing a dynamic flow regime could even result in improving the river wildlife habitat, by halting floods and subsequent erosions over wet periods, while also preventing extended droughts. This concept should ideally be holistic, but more importantly should rely on valid data, and produce verifiable results, taking into account the needs of the local communities.

Care must also be taken that any flow scenarios produced are within existing guidelines, and can be reasonably followed by a dam operator without sacrificing precious resources. As a dam must coexist with the local environment, so too must our understanding of the environmental flow assessment concept always adapt to an ever-changing world.

Understanding flood-risk intervention

The Southwest Delta of Bangladesh has always been vulnerable to flooding – a challenge that the government is seeking to address in multiple ways

SDGs: 9, 13

Located in the floodplains of the Ganges-Brahmaputra-Meghna (GBM) river basin, Bangladesh is vulnerable to flooding, and the Southwest Delta region of the country is subject to multiple flood hazards.

"Bangladesh has achieved food self-sufficiency and the economy is gradually transforming from an agrarian base towards a modern manufacturing and services economy," says AHM Mustafa Kamal, Minister of Planning for the Bangladeshi government. "Making this growth sustainable is challenging in the face of extreme adverse climate variability, with frequent storm and tidal surges, flooding, and droughts."

While the Bangladeshi government has undertaken several initiatives to achieve the Sustainable Development Goals (SDGs), ranging from national-level policies to local measures, persistent flooding problems inhibit the long-term development prospects of the country.

COASTAL EMBANKMENT PROJECT

There is a lack of reliable scientific information on complex coastal systems of Bangladesh, which is needed by decision-makers for flood-risk mitigation. The Bangladeshi government invested in a major Coastal Embankment Project between the 1960s and 1980s to construct a total of 139 polders (enclosed coastal embankments), to prevent the inundation of agricultural lands by saltwater during surge tides and cyclones. Researchers from the University of Oxford have shown that the construction of these polders had both beneficial and harmful impacts on the extent of inundation caused by different types of flooding. "While polders have protected against storm surges and moderate fluvio-tidal events, they increased pluvial flooding and made flooding more likely during the most extreme storm surges," says Mohammed SG Adnan of the Environmental Change Institute at the University of Oxford.

The polder system had a profound influence on the geomorphology of the coastal zone, as well as contributing to a transformation in human settlement patterns. The region has experienced major land-use changes over the past half-century, following the construction of polders. These changes transformed agricultural lands into aquaculture land, used for shrimp cultivation. As well as increasing flood risk (due to a lack of risk-oriented development), such land transformation could lead to a loss of ecosystem, the forced migration of people, and a poverty trap for locals.

The Bangladeshi government has adopted various policies and strategies over the years in response to these complex problems related to polders. Aiming at the sustainable development of the complex delta region, the Bangladesh Delta Plan (BDP) 2100 has recently been approved by the Bangladesh Planning Commission. Due to complex coastal flooding and salinity intrusions, managing the coastal zone and polders is the plan's major focus.

The government has also rehabilitated 17 coastal polders with financial support from the World Bank. An option of temporary "de-poldering", also known as Tidal River Management, has been tested in some low-lying areas to help them partially recover from such problems as the lack of sediment deposition, land subsidence, channel siltation and salinity intrusion.

Despite these various flood-risk interventions, further choices will need to be made in the face of rising sea levels. The world needs a better scientific understanding of the impact of different coastal flood adaptation options in a complex system. Therefore, a new approach is required to help assess potential flood interventions and existing measures before and after their implementations. It is also important to understand the biophysical components of any interventions, and the societal impacts these may have.

Glasgow flood prevention

The White Cart Water Flood Prevention Scheme has helped protect Scotland's biggest city from costly and damaging catastrophe

SDGs: 11, 13

OPPOSITE

Hydro-Brake flow-control devices were key to the operation of the scheme's storage areas

The White Cart Water Flood Prevention Scheme is designed to protect 1,750 properties and businesses in the south of Glasgow from the risk of flooding and more than £100 million in flood damages. The White Cart Water is a shallow, fast-flowing river that is prone to flash-flooding – water levels can rise by 6 metres after only 12 hours of rain. Since 1908 it has caused more than 20 serious floods throughout Glasgow's south side, and in January 1984 more than 500 homes were devastated by flooding.

Upstream storage holds back water during storm events, enabling downstream urban flood defences along the river to be reduced in height and length. A flood-alleviation scheme was based on a holistic catchment management principle of looking for a solution that stretched beyond the city boundaries – as water doesn't care about city boundaries. Three rural flood-storage areas at Blackhouse (Earn Water), Kirkland Bridge (White Cart Water) and Kittoch Bridge (Kittoch Water) were constructed. The scheme achieved Excellent ratings in CEEQUAL for its sustainability performance and was highly commended for landscape and community relations in the CEEQUAL Outstanding Achievement Awards.

Initially, the focus was on identifying potential sites for flood storage areas in the upper parts of the catchment. Overall, 33 sites were considered for size, topography and geotechnical suitability to allow a dam up to 15 metres high to be constructed and the associated environmental impacts. To significantly reduce the lengths and heights of flood defences downstream, it was established that at least three flood-storage areas would be required upstream. This had to sufficiently constrain the flow of water on the White Cart Water and its two major tributaries, the Earn Water and Kittoch Water, with a capability to hold back more than 2.6 million cubic metres of flood water.

FLOW CONTROL

Central to the successful operation of the storage areas was the installation of the world's largest Hydro-Brake flow-control devices into the three dams at each storage area. The Hydro-Brake's internal geometry is designed to enable water to flow unrestricted through it for as long as possible. A self-activating vortex is created when the water upstream reaches a pre-determined height in a flood situation, throttling back the water, and releasing it at a controlled rate. The flow controls have a much larger waterway area when compared to an orifice flow control and this reduces the frequency and duration of flooding at each storage area by allowing higher flows to pass forward at low water depths.

The second element of the project involved the construction of flood defences downstream in the city. This entailed the design and construction of 6km of walls in selected parts of the river corridor downstream, the raising of two footbridges and the construction of eight surface-water pumping stations.

It was important to avoid severing any publicly accessible parts of the river with high flood fence walls, and this was achieved by having an average wall height of 0.85 metres. Particular attention was paid to the alignment of the flood defence wall; where possible, it was sought to maximise the retention of the natural flood plain and follow existing boundary walls to avoid reducing private gardens. These things fed into the design philosophy.

On 29 November 2011, the scheme passed its toughest test yet when it held back almost 240 million gallons of water after a month's rain fell in a day. Around £11 million of damage was averted when flood water was spread over the naturally contoured storage areas before being released in a controlled fashion over the following days.

A smart flooding solution

An eco-friendly flood-prevention scheme that makes use of innovative technology has provided Leeds with a host of benefits

SDGs: 13, 14, 15

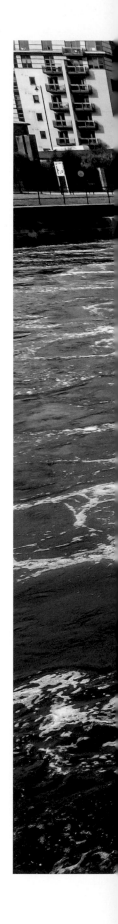

The Leeds Flood Alleviation Scheme is one of the largest river flood defence projects in the country and has helped reduce the risk of devastating physical and economic damage from flooding, while supporting infrastructure development and encouraging investment.

"A key aim of the project was to identify and build in sustainable construction techniques," says Dan Whiteley, Head of Environment at the contractors, BAM Nuttall. "By replacing Victorian weirs with innovative mechanical weirs, we've put Leeds at the forefront of national flood defence schemes."

The scheme was designed to defend Leeds city centre against 1 in 75 year river floods. However, through optimisation, this was increased to 1 in 100 years with allowance for climate change to 2069. Measures were added to reduce the risk of surface-water flooding and to increase the resilience of an emergency response, helping people live and work safely in the city. This more holistic solution included community ownership and increased local awareness of flood protection measures.

CUTTING-EDGE TECH

Moveable weirs were installed to replace existing fixed Victorian weirs at two locations on the River Aire. This pioneering technology, the first of its kind for a flood-protection scheme in the UK, has attracted both national and international interest, as it represents the cutting edge of flood-defence solutions. Normally, these weirs are kept up to maintain deep water in the channel and to ensure that the watercourse remains navigable for boats. But when needed, when the river is high, the weirs can be lowered, causing the water level to drop.

The new weirs allowed for a significant reduction in the required height of river walls and embankments. Together with public realm and landscaping, this ensures that physical and visual connectivity with the waterfront is maintained.

Early in the design phase, workshops were carried out to reduce waste and maximise resource efficiency. From these, reinforced plastic hand railings were chosen in place of traditional cast iron ones, and a recycled gas pipeline was used in place of concrete CFA

(continuous flight auger) piles. These efficiencies, among others, helped significantly reduce the carbon footprint of the project.

All materials used had to be sympathetic to the tone and character of the area, and across the project, the design was interrogated to use existing structures and earthworks embankments instead of building new flood defences. The construction of linear defences such as low-level embankments and riverside walls was designed to blend in with the local architecture.

The team removed a 600-metre man-made island ,which separated the river and an adjacent canal. This was integral to the scheme, and increased river capacity to take the flow of water when the moveable weirs are lowered, essentially removing flood water from the city centre. Some of the excavated material was reused in the diversion of the Trans-Pennine Trail, and the remaining material was used at a local remediation and development project, diverting all material from landfill. This diverted section improves the natural environment and includes a new bridge over the weir to create an iconic gateway into Leeds.

More than 150 direct jobs were created and 22,000 more were safeguarded during this project, and the flood risk to over 3,500 residential and commercial city centre properties was radically reduced. These benefits, together with the protection of key access routes and infrastructure, has helped support the growth and regeneration of the Leeds economy. The South Bank of the city required particular emphasis as one of Europe's largest regeneration areas. The project helped create 35,000 future jobs and 4,000 new homes.

In addition, the previous stone weir's height of three metres formed a barrier to fish moving up the river. The project improved the local ecology with enhancements such as the planting of over 700 new trees; the creation of fish-spawning habitats; and new fish, elver and otter passes at both of the new weirs. These measures help improve species migration.

As well as improving accessibility and water quality, there has also been increased otter activity on the river. Indeed, salmon have even been spotted through the city centre for the first time in more than 200 years.

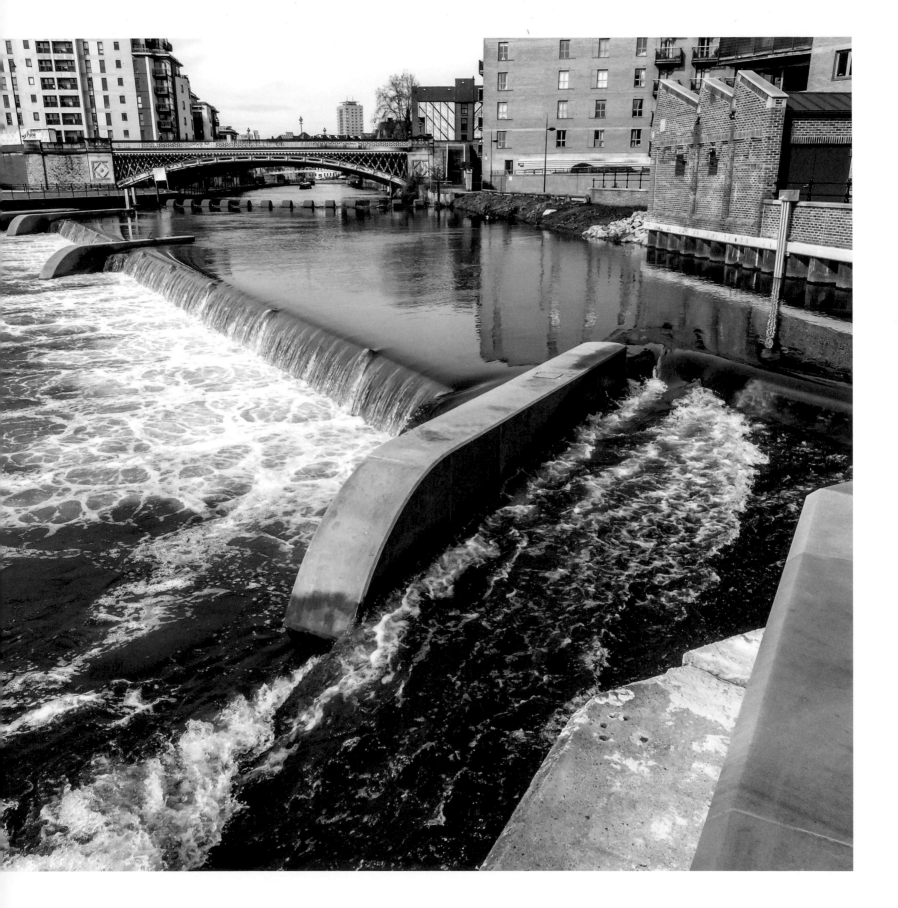

ABOVE
The flood-prevention scheme has
benefitted the local ecology and
economy alike

HEALTH, WATER AND FOOD **93**

Change
for good

For nearly 40 years, WaterAid has been transforming lives around the world with clean water, good sanitation and sustainable solutions to the water crisis

Clean water, decent toilets and good hygiene are basic human rights. They should be a normal part of daily life for everyone – but for millions they aren't. Today, one in ten people around the world still do not have clean water close to their home. And 2 billion people – almost one in four of the global population – do not have a decent toilet of their own.

The world's water crisis is getting worse. Globally we use six times as much water today as we did 100 years ago, driven by population growth and changes in diet and consumer habits. People living in extreme poverty and hard-to-reach communities are often overlooked while others thrive. This lack of basic services can ruin people's lives and undermines the possibility of sustainable development. Having clean water, sanitation and good hygiene can help to transform people's lives.

WaterAid started out in 1981, when the UK water industry came together to create the world's first international organisation dedicated to the water crisis. Since then, it has remained resolutely focused on providing access to clean water, decent toilets and good hygiene for everyone, everywhere by 2030.

This international not-for-profit organisation works with partners to transform millions of lives every year, operating in 28 countries across Africa, Asia, Central America and the Pacific. Offices in the United Kingdom, the United States, Australia, Sweden, Canada, Japan and India – backed by 400,000 supporters – help make this possible.

As the world's largest civil-society organisation focused solely on water, sanitation and hygiene, WaterAid has built a deep understanding of the issues that face communities living without these basic services and of the solutions needed. It starts with taps and toilets – and it goes much further, from educating people on the importance of good hygiene, to changing policy at government level. WaterAid's rights-based approach sparks chain reactions that shift people's habits and expectations. It transforms how countries, not just villages and communities, work and delivers lasting change in people's lives.

BUILDING RESILIENT WATER SERVICES
The World Bank estimates that, by 2025, about 1.8 billion people will live in regions or countries without enough water. To build the resilience of these communities, there is no alternative to urgent and decisive action to build strong water services that can adapt to and manage the impacts of climate change.

It is not enough to simply deliver services and train people how to operate them. Institutions must be in place to help people keep their services running over time and adapt to changes in the water resources available to them. This requires expertise in service delivery and water resource management.

Following the water crisis in Maputo in 2018, triggered by a three-year drought, WaterAid convened the water, sanitation, development and business sectors to develop a long-term strategy to bring water security to Mozambique's capital city. Since then, WaterAid has been working with a dedicated task-team to develop an ambitious programme that addresses the needs of the poorest people, strengthens the existing infrastructure for clean water, toilets and hygiene, and builds ongoing water security. Together, this will provide a firm foundation for Maputo into the future.

WaterAid works hard to reduce the level of water-related risks people face, which force the world's poorest communities to drink dirty water. Many of these risks can be managed, but often the political will and institutional capacity does not exist to make this happen. As a result, hundreds of millions of people remain without access to this basic human right.

ADAPTING TO A CHANGING CLIMATE
The changing climate is making life harder for the world's poorest people, who are already struggling to get clean water. Those who tread lightest on the planet and contribute least to rising global temperatures are carrying the heaviest burden.

WaterAid recognises that to withstand the impact of climate change requires game-changing solutions conceived and delivered through partnerships between public, private and philanthropic sectors, as well as civil society and affected communities.

The organisation works directly with affected communities to build their capacity and resilience. For example, it trains local people to measure rainfall and water-table levels so that communities can recognise when water supplies may be running low. This enables them to make decisions that will help prolong the supply during the dry season.

WaterAid plays a critical role in showing water as a solution to the challenges of adapting to a changing climate. The need to

address this issue has never been greater but, with significant action, change is possible. Millions of people living in extreme poverty are being failed by systems that cannot ensure sustainable, affordable and accessible water and sanitation services. It is a reality that hand pumps break down, toilet pits fill up, and good hygiene habits are forgotten. So, to ensure that services are maintained and will keep working into the future, WaterAid works in partnership to develop robust governance, financial, institutional, environmental and technical capacity. It advocates for systems that will be sustainable over the long term.

To make a lasting difference, clean water, decent toilets and good hygiene must remain a part of daily life long after they are introduced. It's crucial that everyone plays their part to help solve the global water crisis.

One example of WaterAid's work is in the village of Moramanga in Madagascar, not far from the island's capital city of Antananarivo.

The village used to have no clean water, no toilets and no soap for handwashing. Rather than turn on a tap, one local girl, Cynthia, had to walk to a waterhole every day with her mother. This water was dirty and made them ill, but they had no other choice. It was impossible to stay healthy in their village, making it hard for adults to earn a living or for children to go to school. But, with help from WaterAid and its partners, Cynthia's neighbours installed clean water points, built toilets and started to practise good hygiene.

The water comes from a gravity-fed water system – a sustainable water supply that doesn't require specific treatment. Trained local technicians are in charge of its maintenance. Cynthia's village now has four community water points and one new toilet block for the primary school. With the health and freedom these three simple human rights bring, life in this small village is changing for good.

ABOVE
Classmates wash their hands at school in Moramanga, Madagasgar

OPPOSITE
A mother and her two daughters in Cambodia, where 3.8 million people living in rural areas have no access to clean water

"WaterAid's rights-based approach sparks chain reactions that shift people's habits and expectations. It transforms how countries, not just villages and communities, work and delivers lasting change in people's lives"

CREATING LASTING CHANGE

In 2015, world leaders came together to set the UN Sustainable Development Goals (SDGs). WaterAid, together with other water, sanitation and hygiene (WASH) advocates around the world, celebrated the inclusion of a specific goal to ensure availability and sustainable management of water and sanitation for all by 2030 – SDG 6.

SDG 6 is one of the most interconnected goals – improved access to WASH is a basic human right, giving people in remote communities an equal chance to be healthy, educated and financially secure. But, at current rates of progress, everyone in the least developed countries won't have safely managed water until 2131 – more than 100 years behind schedule.

Transformative change is required to build the future we want to see – a future based on reliable access to clean water, where communities have the skills to keep their water flowing. WaterAid works at every level – with governments, businesses and communities – to bring about change for the hardest to reach, recognising their right to clean water, decent toilets and good hygiene.

Today, it continues to highlight the global water and sanitation crisis, and calls for urgent action to reach everyone, everywhere by 2030. WaterAid is determined to make that happen.

Working together to deliver results

ANGLIAN WATER
Location: Petersborough, Cambridgeshire, UK
www.anglianwater.co.uk

In the unlikely location of Peterborough, a quiet revolution in collaborative working is transforming the water industry. It's led by Anglian Water, whose innovative alliance structure shatters the usual piecemeal supplier–client relationship by bringing partners from different companies together under one roof working in genuinely equitable long-term partnerships to deliver the best outcomes for water supply and water recycling.

This policy of total and long-term integration extends to the way Anglian Water organises everything from water distribution and supply to maintenance, water recycling, IT, apprenticeships and even the company's charitable endeavours with Water Aid in Nepal. The results are telling.

"When we had the 'Beast from the East' a few years ago we were able to call upon all the alliances we had developed," says Andrew Page, the Head of Alliance Contracts Management. "We could divert resources from one element of the alliance to the other, so the people who would normally install water meters were able to fix burst mains, without any of them worrying about their contracts. We were singled out for praise as there was almost no disruption to our customers."

A vital strand in the Anglian success story is the company's commitment to carbon reduction – aiming to reach carbon net zero by 2030. Strict targets have been set and are already being broken. But these require innovative and groundbreaking solutions. These solutions also save money, as embodied carbon is expensive. Some of these ideas are brought directly to Anglian at the company's Innovation Shop Window in Newmarket. "People from start-ups and large companies are welcome to present products and theories that might improve water supply," says Page. "One of the big successes that came out was leak detection, where we started using the sort of technology the US Navy uses to detect submarines. We then share these ideas across industry, not just utilities."

When Anglian recently completed a five-year project to improve Grafham Water Treatment Works near Huntingdon, the project's carbon footprint was reduced by 50 per cent with a financial saving of £30 million. This was partly down to the extensive use of digital engineering and design including four-dimensional modelling of construction and the creation of walk-through models to discuss design changes and identify any potential clashes with existing operations.

"That is a great story about a big bit of infrastructure that we found a different way of doing that didn't need the more traditional solution," says Page. "There are different approaches that have the same effect – lower carbon means lower cost. Many of our schemes are financed by Green Bonds. We were the first utility company to issue a Green Bond to help finance our capital build programme. Every scheme we undertake meets strict environmental criteria and carbon-recudtion targets."

This strongly supports to Anglian's Public Interest Commitment to ensure its work positively impacts on the wider environment and the local communities it serves. Commitment to social purpose is embedded in the company articles, making them legally bound to consider environmental and community impact – a triple bottom line. Anglian's alliances are involved in this way of thinking too. One supports a community farm in Wisbech that helps people with mental health issues and learning difficulties. Meanwhile environmental benefits were uppermost at the water company's recent scheme at Ingoldisthorpe in North Norfolk. The brief was to improve the quality of the effluent coming out of a water recycling centre. Instead of investing in more carbon-hungry infrastructure and chemical treatment, it was decided to create a wetland to be designed, built and managed by Norfolk Rivers Trust with funding from Anglian Water. The wetland provides natural filtration, removing ammonia and phosphate and converting it into energy for plants before the water reaches the river.

"We have good biodiversity on this three-hectare site and didn't need to pour any concrete," says Chris Gerrard, Biodiversity Manager at Anglian. "It's good for the environmental footprint, it's putting wildlife into the valley and it's raising the water quality.

It's win, win, win all round." Anglian is now exploring other sites in the region where wetlands might be suitable.

Much further afield, Anglian is making a difference in Nepal. Working in conjunction with Water Aid, Anglian is bringing the full strength of the skills and knowledge of its alliance network to the town of Lahan in the south-east of the country. The Beacon Project is a 13-year collaboration between Anglian Water and Water Aid that will see Anglian's alliance partners working directly on the ground to improve facilities in an area where half the population of nearly 100,000 has no access to fresh water.

"We have agreed for the stakeholders to work together, pooling all our knowledge and resources to improve water quality, maximise the infrastructure and develop a strategy that is based on the communities' needs rather than on western assumptions," says David Ward, who is leading the project. "In addition to our partners' considerable charitable fundraising contributions, they are supporting philanthropically with skilled and experienced employees. There are

many layers of opportunity that this approach creates that you don't get in traditional fundraising. Partners are able to see and contribute directly in the areas where the aid is needed, while providing valuable opportunities for apprentices or graduate trainees, helping them build valuable skills and experience for the future."

Some of those graduates will have come through courses sponsored by Anglian at four colleges in the region. This project began in Wisbech in 2015 as part of the Wisbech 2020 vision to raise aspirations and create pathways into employment. Courses in engineering and construction are delivered by Anglian's alliances, who offer site visits and work experience and provide specialist lectures. This allows students to see what challenges the water industry faces but also gives them access to myriad opportunities

provided by the multiple partner companies in each alliance. Around 85 per cent of students continue on to apprenticeship schemes. Apprentices are shared across the alliances, so gain a wide range of experiences.

"At the end of their apprenticeship they will have spent time across all the companies so they are multi-skilled, with the ability to work in any alliance and a full knowledge of the challenges facing the water industry," says Sarah Charman, the Collaborative Skills Coordinator. "The only question then is 'where do they want to work?'" It's a further illustration of the imaginative, joined-up way that Anglian is approaching the pressing issues facing the water industry today, creating a new generation of workers who are ready to work together to make a real difference to our planet.

A revolutionary approach to the water cycle

NMCN
Location: Sutton-in-Ashfield, Nottinghamshire, UK
www.nmcn.com

Engineering has always thrived at that thrilling nexus between challenge and innovation. Engineering and construction company NMCN is one of the UK's largest contractors working within the water sector and it is also among the most creative, having recognised that challenges in water delivery cannot always be addressed by traditional methods.

The company works with a dozen water companies around the UK to design and construct reservoirs, treatment plants and tunnels, but its most groundbreaking work comes with an imaginative approach to off-site construction and phosphate removal. "We are the biggest UK contractor in the water sector in terms of delivering infrastructure both below ground with pipes and sewers and above ground with water treatment plants," says Mark Hanrahan, Business Development Director. "But what matters is what we are doing differently to make a positive impact on people's lives."

NMCN (formerly North Midland Construction and Nomenca) has a commitment to difference that can be seen in its approach to employment. The company has a focus on apprenticeships, graduates and training programmes. There are a far higher proportion of graduates in the workforce than the industry standard and NMCN maintained that position throughout the post-2008 downturn. This "people-centred" approach is what originally led to the company's decision to move more of its work off site into factories.

"We were doing a lot of work on major construction sites and implemented a variety of safety measures but the potential for injury was still very high," says Hanrahan. "We began to wonder whether we could do some of this work in a factory and that's what we did. We employed the same principles but we did it in a safer environment."

It soon became apparent that, as well as improving conditions for employees, a move towards factory builds had several other positive repercussions. For a start, there were no constraints from the weather. It meant equipment could be tested before installation. When work was being done to improve or extend existing facilities, there was no danger of negatively effecting existing operations or worrying about access. All these factors had a positive impact on cost. "We immediately saw the benefits and so did the clients," says Hanrahan.

At Uttlesford Bridge, a major treatment works in Saffron Walden in Essex, NMCN was able to take things to the next level. Affinity Water wanted to improve the efficiency of the plant while ensuring that it remained operational. A new 40-ton component needed to be installed. For NMCN, the solution was an off-site build and the way it went about it illustrates the potential of off-site building – and also NMCN's ambition.

"We laser-scanned the entire site," explains Hanrahan. "We used ground-penetrating radar to detect buried surfaces. Then within the course of two hours we stitched what is called a point cloud together using photographic imagery to create a virtual-engineering environment. Using an ocular 3D headset we could then immerse the customer into the site so they could walk around it. Within two weeks we superimposed a model of the new plant we were going to build, showing precisely how we could integrate it with the existing works."

The benefits were manifold. A project that would have taken years to execute on site was reduced to a matter of months, reducing the work time on site by some 34,000 man-hours. Because the new component had been constructed, assembled and tested in the factory, it was defect free. The virtual element allowed any potential pitfalls to be anticipated and dealt with

> *"These are issues that will shape our future direction and effect the entire industry, we have shown we can face these challenges"*

ahead of construction. The capital saving to the client amounted to 32 per cent and there was also a significant reduction in embedded carbon as less concrete was used. The component was delivered in two parts and was then assembled on site, a process that took two days and required the use of two trucks and a crane. This eliminated an estimated 24 journeys by heavy goods vehicles, which would otherwise have had a negative impact on the environment and local community.

"We knew it was possible and in terms of what we put forward and what was delivered, it was a phenomenal success. Affinity has said that with any future scheme it wants to see what can be done off site," says Hanrahan. "For this to be a success, you need everybody to be fully engaged. Not just the chief executive but every level of the client's engineering team."

One big challenge facing the water industry that NMCN has tackled relates to phosphate removal. In 2018, the government set out a Water Industry National Environment Programme (WINEP) in the form of a set of environmental targets to be achieved by 2025. One related to the removal of phosphate, which overstimulates the growth of algae and leads to a gradual depletion of oxygen, creating lifeless waterways. For the very smallest treatment plants where there is not the capacity or footprint for other methods, phosphate removal is usually done with chemical dosing. "Dosing needs to take place at around 800 sites in the UK over the next five years, but there are insufficient chemicals to treat all those sites," says Hanrahan.

Keeping this firmly in mind, NMCN has developed an alternative to chemical dosing in the form of a sono-electrical chemical treatment, which combines electrolysis and ultrasound to generate a reactive treatment. "You do away with any need to use ferric sulphate," says Hanrahan. "It has a smaller footprint, uses less energy and you don't need to rely on the delivery of chemicals. It is ideal for the smaller sites that provide us with the biggest challenges."

With its commitment to off-site construction and phosphate removal, NMCN has illustrated that it is not afraid to overcome the hurdles facing the water industry. "These are issues that will shape our future direction and effect the entire industry, and we have shown we are doing what we can to face these challenges," says Hanrahan. "Innovation is essential to ensure we protect the environment and deliver better solutions. We can't just dig holes and pour concrete."

The full potential of wastewater

BIOWATER TECHNOLOGY AS
Location: Tonsberg, Norway
www.biowatertechnology.com

Water pollution is a global issue, and that is why a small Norwegian company comprising renowned experts in its field has delivered more than 80 treatment projects in dozens of countries across almost every continent on the planet.

Biowater Technology AS works at the cutting edge of biological wastewater treatment, providing smart energy-efficient and environmentally friendly solutions to the worldwide problem of sewage and industrial wastewater. "The vast majority of wastewater is treated unsustainably in both the processes used, and in the environmental impact of the effluent discharged," says the company's CEO Ilya Savva. "We want to ensure that both clean water and the environment are absolutely protected by implementation of best available technologies."

Biological wastewater treatment uses naturally occurring micro-organisms such as bacteria to clean polluted water. Biowater works with local municipalities and industries globally to beat targets

for wastewater treatment. "These sectors are increasingly motivated to reduce their discharge and save energy, and are also realising the energy potential of wastewater," explains Savva. "We see the potential of what can be achieved and offer numerous ways to help customers."

One recent wastewater treatment plant at a food and beverage factory in Grimstad in Norway is an example of how Biowater operates. Using patented technology, the plant is sustainable and produces minimal sludge, treated wastewater with great potential for reuse, and renewable energy in the form of methane. Around 75 per cent of the organic waste is turned directly into methane gas, an energy source generated at the point of use which is far more environmentally sustainable than extracted, refined and transported fossil fuels. This combination of a renewable energy-generation anaerobic stage and aerated biofilm treatment in a single reactor is unique to Biowater.

"Our patented techniques are continuous enhancements of proven technologies, only achievable through our experience of working in the industry for so long," explains Savva. "We create more energy-efficient versions of the existing standard processes – more sustainable in terms of energy, waste production and footprint. What makes Biowater stand out is that we understand

there are different biological treatment techniques for a full range of wastewaters by sector and by industry, and our team can tailor treatment solutions for each individual customer and their needs. We design the best process and biological treatment solution for each and every client , and we are proud that our patented technologies are increasingly being selected."

For this to work effectively, it is essential to build strong partnerships. "We want to understand our partners' needs and give them positive options so we deliver the best solution together," says Savva. "And we want to support them so they can really understand our technology, its strengths and how to use it most effectively."

Biowater processes are now recognised throughout the world. Savva has worked in the civil engineering industry for his entire career and recognises wastewater treatment being taken more seriously, but, in his view, not seriously enough. "We are actively working towards solving a global crisis in terms of preserving natural resources and saving the environment," he says. "And we do this by using our knowledge and experience to enable people to find solutions. At Biowater we all believe in our mission – we believe in it passionately. This is a very exciting and fast-changing industry, one that is driven by a very real need."

A firm focus on sustainability

DUPONT
Location: Switzerland
www.dupontwatersolutions.com

Despite, or perhaps because of, its 200-year history of innovation, a new culture has emerged at DuPont. A world leader in specialty products, in 2017 the firm merged with Dow, resulting in the formation of three distinct companies – Dow, Corteva and DuPont – each with a new and specific focus. The new DuPont has its sights set on becoming a global force in sustainable innovation and in October 2019 announced nine objectives inspired and guided by the UN's Sustainable Development Goals (SDGs).

"Water is central to these goals, mainly impacting SDG 6," explains Alexander Lane, DuPont's Commercial Director. "It is fundamental to all human ventures and survival, and we have been challenged to grow this sector substantially and quickly."

The new company structure will help achieve this, with DuPont's new emphasis enabling the water sector of the business to take a more prominent role. "We are now a large business in our own right rather than a smaller part of a global petrochemicals giant," explains Lane. "That's good for us because we earn more focus and we are very interested to grow faster in this space. We are investing and making acquisitions, which is very exciting for us. We are hitting current targets, we've invested a great deal in existing assets so they can be upgraded, and we'll be doing more to make our customers happier."

Water scarcity is one of the target areas. The World Business Council for Sustainable Development (WBCSD) predicts that water demand will expand hugely in the next 30 years – up to 70 per cent in the municipal/industrial sector and 85 per cent in the energy sector. To address this issue, DuPont is working on ways to make wastewater reusable through advanced technology and has become a world leader in water purification and specialty separation technologies. The company has acquired two businesses that specialise in ultrafiltration technology: BASF's ultrafiltration membrane business "inge" and the Memcor business including ultrafiltration and membrane biofiltration technologies from Evoqua Water Technologies Corp. In addition, DuPont acquired Desalitech Ltd, a closed-circuit reverse osmosis (CCRO) technology

provider, and OxyMem Ltd Membrane Aerated Biofilm Reactor (MABR) technology for the treatment and purification of municipal and industrial wastewater.

As an additional element of its new sustainability commitments, DuPont will align 100 per cent of its innovation portfolio with the UN's SDGs, meaning it will only support the research and development of projects that meaningfully advance the UN's targets. DuPont has promised to design all of its products and processes by using sustainability criteria. The company will reduce greenhouse gas emissions by 30 per cent and draw 60 per cent of its electricity from renewable energy sources, and by 2050, it intends to be carbon neutral.

"We want to be bigger, bolder and better," says Lane. "DuPont has a strong history of market-based innovation and our contribution to the UN's sustainability goals is helping to drive growth and purpose. Everything DuPont does should touch these goals where they are relevant to us. But getting there isn't a journey we alone control, we need other people, and that's where we look at SDG 17, which concerns the value of establishing and forging new partnerships. This fresh perspective has enabled us to go and talk to different people."

These conversations have led to pioneering projects in Kenya and Egypt, countries where parts of the community still do not have daily access to clean, safe water. For example, Kasarani is in Nairobi, where the groundwater contains dangerously high levels of fluoride, capable of causing dental and skeletal fluorosis. To ensure the district's St Francis Hospital Training Centre has access to safe drinking water, a filtration system was installed by a collaborative group that included DuPont Water Solutions, USAID and local original equipment manufacturer Davis & Shirtliff.

"Most of the water in East Africa is sourced from boreholes and in certain areas it's very high in fluoride, which is good in small portions but can lead to complications in higher doses," says Lane. "We worked with a small community to install a small but effective system using nanotechnology to make sure that we could cleanse the water for the school and hospital. It was originally intended

"We want to be bigger, bolder and better. Our contribution to the UN's sustainability goals is helping to drive growth and purpose"

to impact the lives of 6,000 people but word has spread and people are coming from tens of kilometres because they understand that this water is safe to drink. The project has fostered further innovation too, with the installation of preloaded credit-based automated dispensing systems: safe water can now be collected outside typical office hours so it's more freely available to the general community."

The project was only possible because of the input of several different parties, including NGOs and government agencies. As a result, water-borne disease has been greatly reduced, something that has also happened in Egypt, where DuPont helped construct a compact treatment plant to serve four villages around Suez. This system delivers high-quality drinking water without the need for treatment chemicals. It is built from pre-engineered and pre-assembled ultrafiltration units with a modular system design requiring fewer parts and materials than traditional designs, resulting in a simplified installation process. The facility was fully operational in just three months rather than the usual 12. Further modules can be retrofitted to increase capacity by 25 per cent bigger without any mechanical or electrical interventions and it is self-contained – transportable to different areas as needed.

"Decentralised communities need a decentralised water supply," says Lane. "In Egypt they understand that if water can be providing clean water to a community increases the health of a population and reduces medical costs, which benefits society. People don't need to walk long distances to fetch water – which is often the job of women and girls who can miss out on schooling as a result. The system we installed is a containerised solution designed with our partners. It can be located wherever needed, plugged into water and power sources and can immediately start producing clean water. Egypt is anticipating greater water shortages in the future; these solutions are needed as water becomes increasingly scarce and precious."

The new ultrafiltration plant has been celebrated for its effectiveness and agility. It was named Best Disruptive Technology at the 2019 International Desalination Association Awards. It's a perfect example of where DuPont wants to be as the company contemplates a bright future constructed on the solid foundations created by existing personnel. "We are seeking to build further on our existing deep expertise and passion," says Lane. "We've undergone a rejuvenation. While there's still much to do, there's a lot of enthusiasm for the journey ahead. We strive to continually improve our offering and our service – and that will be good for our customers and good for us, too."

Setting the standard in forward thinking

SOUTHERN WATER
Location: Worthing, West Sussex, UK
www.southernwater.co.uk

As a company that deals directly with one of the world's most precious assets, Southern Water fully appreciates the importance of sustainability when it comes to water. The company operates across Sussex, Kent, Hampshire and the Isle of Wight, supplying 2.5 million customers with clean water through 53,737 km of pipework and via an infrastructure framework that encompasses 205 service reservoirs, 84 water treatment works and 365 wastewater treatment works.

Its engineers abide by the ICE charter for sustainable development, which states that: "The Institution of Civil Engineers believes that Sustainable Development is central to civil engineering and that ICE and the profession it serves must organise themselves accordingly". Southern Water's work crosses over with the UN's Sustainable Development Goals in several areas – SDG 6, access to clean water and sanitation; SDG 8, creating decent work and economic growth; SDG 9, industry, innovation and infrastructure; SDG 11, sustainable cities and communities; SDG 12, responsible production and consumption; and SDG 14, life below water.

Southern Water has set its own self-defined standards to reinforce the public industry commitment established by the English water industry in 2019. The latter includes achieving net zero carbon emissions by 2030, tripling the rate of leakage reduction, making bills affordable to end water poverty and preventing the equivalent of 4 billion plastic bottles ending up as waste. Southern Water does not stop there. "We want to set the standard for best practice in the industry," explains Enabling Manager Dr Nicola Meakins.

Southern Water's own five transformation plans are heavily linked to sustainability. These include the commitment to reduce the average daily consumption of water to 100 litres per person by 2040. The company also wants to ensure that water catchment is at the heart of decision making and delivery; create a more resilient supply network; transform wastewater treatment works into community assets; and create drainage capacity across the sewer network. Numerous things need to be done for these to be achieved, such as creating a sustainable urban drainage system that is better able to store water or discharge it direct into the ground. There is also the need to build more cleverly and efficiently, embrace new construction technologies, eliminate waste, minimise resource consumption, streamline delivery process and design for low-carbon materials.

Evidence of the success of Southern Water's strategy can be seen in wastewater treatment plants at Peacehaven in East Sussex and Woolston in Hampshire. The former treats 95 million litres of water per day to ensure sewage isn't discharged into the sea close to the coastline. It is located in an area of outstanding natural beauty so environmental studies were carried out ahead of construction, while the plant was built to blend into its surroundings by being hidden in a hollow and covered with what at the time was the largest turf roof in the UK.

Guiding this process is Dr Meakins' team of 20, which includes environmental advisors, ecologists, surveyors and planners. "We are involved in scoping for environmental concerns, de-risking, and the environmental care and protection of our capital delivery schemes," she says. "We have specialists in all aspects of environmental management, including bats, badgers and archaeology — everything we need to ensure that our sites are properly looked after during design and construction. We get all the necessary environmental

> *"We want to go further than we are
> required to. We want to set the standard
> for best practice in the industry"*

consents and licences to undertake the construction work and we have demonstrated to our conservation regulators, Natural England, that we are competent and trustworthy, so they have awarded Southern Water an Organisational Licence which effectively means that in many cases we can self-regulate. This isn't a bolt on or afterthought, it's completely embedded throughout our strategy. Other teams are upstream of the process and we all take part in the planning a long time ahead."

In the Southampton suburb of Woolston, the challenge was to replace an existing wastewater treatment works. This had been built in 1966 and was in drastic need of modernisation. The new treatment work had to reduce unpleasant smells that were leaking from the site, it needed to ensure the treated wastewater met more exacting environmental standards and it had to fit in with the ongoing residential development of the area. The stylish building achieves the last of these so effectively that when Senior Project Manager Stewart Garrett was showing somebody round the finished plant, they assumed it was a state-of-the-art conference centre.

Construction began in 2014 and was completed at the end of 2019. An "Employment and Skills" plan encouraged the participation of local contractors, sub-contractors and suppliers to provide employment opportunities for Southampton residents, while a Community Liaison Group provided a link between Southern Water and the local community with regular meetings. This concern for community was maintained by the design, with the new building having a double skin to ensure unpleasant aromas do not affect the surrounding neighbourhood. The effluence is treated with an MBR

(membrane bio-reactor) capable of treating flows of up to 427 litres per second. It is the largest MBR plant in the UK. "The membrane is effectively a very fine mesh that takes out all the micro-solids," says Garrett. "If you think about traditional wastewater treatment, the membrane replaces the old filtration system. It's quite an expensive process but it will mean we reach the highest standards."

Southern's stringent targets for reducing carbon had an impact on the design and construction of the new plant. "We look at construction holistically and ensure we look at the environmental issues right across the life cycle, often before the engineers get involved," says Meakins. "We have project managers come to us at the very beginning of the process to ask the best line of attack and we get involved throughout the process. We go out on site to make sure things are dealt with, and we are there for the post-construction and handover."

At Woolston, Southern's engineers were especially mindful of building with longevity in mind. That meant ensuring standards for the treated water were exceptionally high, but also that the building itself would last. "We are moving away from concrete structures wherever we can as there is a great cost in concrete, a lot of power and materials are required, and it's not sustainable," says Garrett. "What we are trying to do is create assets that can be built off-site. We are looking at just-in-time delivery, where we construct tanks and structures off-site and deliver them when and where they are needed. Our engineers always consider the life cycle of the asset as well as the material when looking at the best way forward. We want this to be sustainable for future generations so it has been designed to last for many years."

An agency of change in water management

ENERGY & WATER AGENCY
Location: Malta
www.energywateragency.gov.mt

On the Mediterranean island of Malta, where natural water resources are limited and where consumption is increasing due to marked population growth, ensuring water sustainability is a key concern. This has led to the creation of the Energy & Water Agency (EWA), an organisation that supports the government in the formulation of comprehensive water and energy policy.

"Malta's small size provides a unique opportunity for piloting new approaches and adapting technologies to develop a specific water-management framework," says CEO Manuel Sapiano. "Malta's policies recognise that natural resources are not sufficient to meet the growing demand for water. There is no single solution to address these challenges, but instead one needs to identify and implement a diverse but effective portfolio of measures."

EWA's role is ensuring the development of a comprehensive water-management framework. Within this context, Malta's water utility, the Water Services Corporation, has embarked on a project to drastically reduce its net abstraction from the groundwater aquifer systems by increasing water use efficiency, reducing leakages and introducing alternative supplies (desalination and water reclamation).

Furthermore, from a broader environmental perspective, energy needs for water production, distribution and treatment are also being reduced, cutting down impacts from energy-related carbon emissions.

As well as working with pioneering schemes like the "New Water" reclamation project (pictured, above), EWA supports a wide-ranging national water conservation campaign which aims to help Maltese citizens make the correct decisions on using water more sustainably. The campaign emphasises the role each citizen must play to ensure an effective use of water at a national level. This campaign is complemented by a focused and interactive educational programme for students at the National Water Conservation Awareness Centre.

Some of these activities have a scope which extends beyond Malta's shores. The EU-funded LIFE Integrated Project, led by EWA, looks at the development of a sustainable water-management strategy which can be replicated in similar regions. "Through the involvement of regional partners such as the Union for the Mediterranean, who have recognised this project as one of regional importance, we will be seeking to disseminate Malta's water management approach," says Sapiano. "This could have a positive impact on the whole region."

EDUCATION, RIGHTS AND EQUALITY

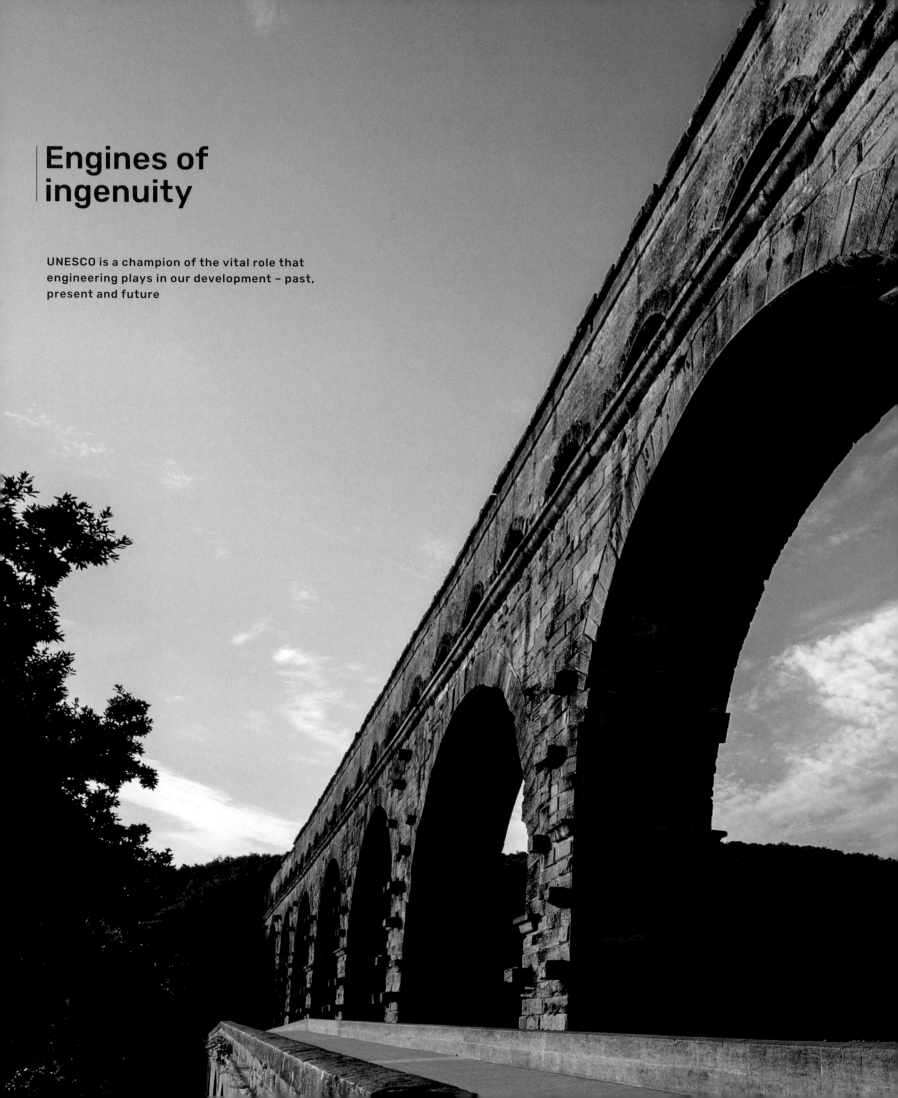

Engines of
ingenuity

UNESCO is a champion of the vital role that
engineering plays in our development – past,
present and future

The Pont du Gard in southern France, the Parthenon in Athens, the Great Wall of China and the Pyramids of Giza at Memphis are just a few of the sites on UNESCO's World Heritage List that inspire awe. Today, these sites serve an aesthetic and historical purpose, but at the time of their construction they were works of engineering that served practical purposes. For example, Roman aqueducts at Pont du Gard or Segovia in Spain provided essential water and sanitation to cities.

UNESCO (the United Nations Educational, Scientific and Cultural Organization) has always paid particular attention to engineering. "Engineering has helped to shape the world for millennia and now more than ever, the world needs engineering," says Audrey Azoulay, Director-General of UNESCO. "Engineering is also needed to help humanity deal with the second major change of our time, that of technological disruption. The tech revolution has given rise to new opportunities, but it also poses new ethical dilemmas in the fields of artificial intelligence and robotics, for example."

Engineering has also had an essential role in achieving UNESCO's mandate and global and strategic priorities. In fact, UNESCO was established during a conference that was held in November 1945 at the London headquarters of the Institution of Civil Engineers, the world's oldest engineering institution. In the early years of UNESCO, engineering was the largest activity, in terms of personnel and budgetary resources, in the Natural Science Sector. During this time, the focus of the programme was on engineering education through human and institutional capacity-building projects.

A DRIVER OF DEVELOPMENT

UNESCO's first major engineering report, "Engineering: Issues, Challenges and Opportunities for Development", published in 2010, defined engineering as "the field or discipline, practice, profession and art that relates to the development, acquisition and application of technical scientific and mathematical knowledge about the understanding, design, development, invention, innovation and the use of materials, machines, structures, systems and processes for specific purposes".

The word "engine" derives from the Latin "*ingenium*" for ingenuity or cleverness. Engineers use scientific knowledge and mathematics to create technologies and infrastructure that address contemporary issues. They connect social needs with appropriate technological innovation and commercial applications. As such, engineering is a major driver for sustainable socio-economic development.

Engineering has always had an essential role in achieving national development and human welfare. All nations have benefited from its findings and practices to reach their targeted growth objectives and desired advancements. UNESCO will soon publish its second engineering report, which will build on the success of the first. It will focus on the role of engineering in development and addressing the UN's 17 Sustainable Development Goals (SDGs).

> *"Engineers are needed to imagine, create and implement the technologies and infrastructures that will make sustainable development possible"*

The second report will make a series of clear recommendations to its key audiences, backed up by the evidence, trends and ideas covered in its chapters and by references to the first report. In addition, the second report will introduce a statistical toolkit to help tackle the challenges of global statistics on engineering. The recommendations and toolkit will emerge from the report's stakeholders and be agreed upon by the Engineering Report II Steering Committee.

UNESCO identifies several key issues. The world is still experiencing, on average, a shortage of engineers in all domains. It is still observing a decline of interest and enrollment in engineering by young people, especially women, and gender parity in the different engineering fields has not been achieved by all countries. The world is still experiencing a brain drain of qualified engineers from developing countries to developed countries. Policy-makers should pay greater attention to the need and importance of engineering for sustainable development. Innovation in engineering is a necessity to better adapt to and address today's global challenges, such as the SDGs.

Engineering is a vital profession in the addressing of basic human needs: in alleviating poverty, in promoting secure and sustainable development, in responding to emergency situations, in reconstructing infrastructure, in bridging the knowledge divide and in promoting intercultural cooperation.

The UNESCO engineering programme was created to promote engineering education at secondary and tertiary levels and to highlight the roles and accomplishments of women and youth in the engineering field. Working with Member States, international partners and programme experts, the programme aims to strengthen engineering education through curricula development and capacity building. The programme is fostering scientific exchange and excellence through its partnerships with various nongovernmental organizations, multinational corporations and engineering educational institutions to encourage investment in applied research and training. In line with UNESCO's Global Priorities, it focuses on gender equality and the African region.

In collaboration with its partners, the UNESCO engineering programme is increasing the visibility of engineering in secondary schools. Through joint programmes and "hands-on" activities, UNESCO and its partners are providing secondary students with the opportunity to explore and cultivate their interests in engineering.

"Engineers are not only expected to find solutions that will help us to adapt to or mitigate the consequences of climate change; they are also needed to imagine, create and implement the technologies and infrastructures that will make sustainable development possible," says Peggy Oti-Boateng, Director of the Division of Science Policy and Capacity Building at UNESCO. "This implies breakthrough ideas in all engineering specialities, whether they are traditional ones, as civil, mechanical or chemical engineering, or emerging ones such as data and artificial intelligence engineering, bioengineering, and even more importantly, fields that we are not even aware of yet."

ABOVE
UNESCO is working
to achieve gender parit
in engineering

EDUCATION, RIGHTS AND EQUALITY   **121**

On course for success

Training programmes devised by UNITAR are helping the developing world's young minds tackle a host of developmental challenges

Far from bragging about it, 25-year-old Hasan might be too modest to acknowledge that he's part of Iraq's best chance for a brighter future. Born in Baghdad one year after the end of the first Gulf War, he grew up in a family of engineers that definitely nurtured his own fascination in science. "This is a field I've always felt a huge interest for, just like some of my brothers and sisters."

In 2010, Hasan entered Baghdad's University of Technology, where he became a specialist in electrical engineering. He also had the opportunity to study at Columbia College and the University of Missouri in USA, and started volunteering for the United Nations Development Programme's innovation for development project as a "Design Thinking" trainer. He obtained his master's degree in 2017 and has since become a trainer at his university, realising one of his dreams.

"My family was very open-minded and supportive, they've always encouraged me to follow my own path," he says. "But I have to admit that, until 2013, I was an ordinary student, with no clear perspective on what that path could be. I just wanted to finish my studies, leave the university and find a good job. Since then, I have become a trainer, but not an ordinary one in the sense that I get to mix real practical life with theoretical knowledge. This is a working method that you seldom encounter in Iraq's academic world.

"During my studies, I was already giving training about entrepreneurship, leadership or team-building to my fellow students, and helped them with how to write a solid CV, how to build an effective communication plan, or how to successfully pass an interview for any kind of job. Then I saw that it could become a full-time job, so I decided to become a professional trainer."

Several big companies in Iraq have sought his expertise, including biopharmaceutical giant AstraZeneca, the country's biggest communication specialist Zain Iraq, and the University of Technology, which has no fewer than 16 engineering departments.

Three different universities, as well as Baghdad's Al Mansour Mall (one of the biggest in the city), have showed interest in allowing him to organize a technology job fair. But it wasn't all smooth sailing, and Hasan still has to deal with some resistance to make his voice heard.

"Getting to work for a university is one of my biggest challenges, because it is so hard to convince people of the importance of allowing students to learn leadership techniques in the engineering field," he says. "Many people in charge in academia have a hard time taking it seriously. Most of the ones I've had to deal with were in their sixties and didn't see that the world they knew had changed."

UNITAR COURSES

When asked about the United Nations Institute for Training and Research (UNITAR) courses he took in 2016, the first word that comes to Hasan's mind is "fascination". "Everything was very well thought out, and not boring at all. The skills taught there were very helpful to me because I already had clear projects in my head but I did not know what were the proper first steps, and I had no idea of the number of things to take into consideration before starting anything. For example, I was especially excited to learn about PESTLE (political, economic, social, technological, legal and environmental) analysis. We were 24 participants of the Fellowship who've become friends since then, and we keep on seeing each other every second Thursdays to talk about our projects."

Hasan is not alone in having benefited from UNITAR's help. Established in 1963 and now based in Geneva, this dedicated training arm of the UN provides capacity development activities and learning solutions to assist developing countries, with special attention to Least Developed Countries (LDCs), Small Island Developing States (SIDS) and other groups and communities that are most vulnerable, including those in conflict situations. More than 130,000 individuals benefitted from the delivery of 671 UNITAR training and related events, in areas as diverse as economic development and social inclusion; environmental sustainability and green development; sustainable peace; and research and technology applications.

Iraqis have gone through tough times in recent years and in the eyes of the younger generation, the future might not look so bright. But Hasan seems immune to any kind of pessimism. "Many friends or people I meet in my daily life are telling me: 'your English is great, you have an academic degree, why don't you leave to work in the United States or somewhere else?' I tell them that I believe in our country and its youth. I am convinced that we can have a brighter future, here and now."

ABOVE
With UNITAR's support, Hasan
has been able to give back to
the society that educated him

Interlinked thinking

The United Nations University's collaborative research contributes to resolving the global problems of human survival, development and welfare that are the concern of the UN and its Member States, as its new report on water and migration exemplifies

Do migrants willingly choose to flee their homes, or is migration the only option available? There is no clear, one-size-fits-all explanation for a decision to migrate – a choice that will be made today by many people worldwide, and by an ever-rising number in years to come because of a lack of access to water, climate disasters, a health crisis and other problems.

Data are scarce on the multiple causes, or "push factors", limiting our understanding of migration. What we can say, though, is that context is everything. The United Nations University (UNU) – a global think tank and postgraduate teaching organisation headquartered in Japan – is among the organisations looking for direct and indirect links between migration and the water crisis. This is a problem that has different faces — unsafe water in many places, chronic flooding or drought in others.

The challenge is separating those push factors from the social, economic and political conditions that contribute to the multi-dimensional realities of vulnerable migrant populations, all of them simply striving for dignity, safety, stability and sustainability in their lives.

A new UNU report, "Water and Migration: A Global Overview", from the UNU Institute for Water, Environment and Health (UNU-INWEH), offers insights into water and migration interlinkages, and suggests how to tackle existing gaps and needs. Its information can be understood easily by stakeholders and proposes ideas for better informed migration-related policymaking, including a three-dimensional framework applicable by scholars and planners at multiple scales and in various settings.

The report also describes some discomforting patterns and trends. By 2050, a combination of water- and climate-driven problems and conflicts will force 1 billion people to migrate, not by choice but as their only option. Links to the climate change and water crises are becoming more evident in a dominant trend: rural-urban migration.

That said, there is a severe lack of quantitative information and understanding regarding direct and indirect water and climate-related drivers of migration, limiting effective management options at local, national, regional and global scales. Global agreements, institutions and policies on migration are concerned mostly

ABOVE
The shrinking of the Aral Sea illustrates
the report's call to recognise the impact
the environment has on migration

126   UN75: SUSTAINABLE ENGINEERING IN ACTION

"There is a severe lack of understanding regarding direct and indirect water and climate-related drivers of migration"

with response mechanisms. What is needed is a balanced approach that addresses water, climate and other environmental drivers of migration.

Unregulated migration can lead to rapid, unplanned and unsustainable settlements and urbanisation, causing pressure on water demand and increasing the health risks and burdens for migrants, as well as hosting states and communities. Migration should be formally recognised as an adaptation strategy for water and climate crises. While it is viewed as a "problem", in fact it forms part of a "solution".

Migration reflects the systemic inequalities and social justice issues pertaining to water rights and climate-change adaptation. Lack of access to water, bad water quality and a lack of support for those impacted by extreme water-related situations constitute barriers to a sustainable future for humankind. Case studies in the report provide concrete examples of the migration consequences in water- and climate-troubled situations. There is the shrinking of Lake Chad in Africa and the Aral Sea in Central Asia; the saga of Honduran refugees; the rapid urbanisation of the Nile delta; and the plight of island nations facing both rising seas and more frequent and more intense extreme weather events.

In addition, the added health burdens imposed on people and communities by water pollution and contamination create vicious cycles of poverty, inequality and forced mobility. While the Sustainable Development Goals (SDGs) agenda does not include an explicit migration target, its mitigation should be considered in the context of SDGs that aim to strengthen capacities related to water, gender, climate and institutions. These issues resonate even as the world deals with the COVID-19 pandemic.

Recent news stories have chronicled the plight of desperate migrant workers trapped in the COVID-19 crisis in India, and of displaced people in refugee camps where social distancing is unachievable, as is access to soap and water, the most basic preventive measure against the disease. Add to that the stigma, discrimination and xenophobia endured by migrants that continue to rise during the pandemic.

Even at this moment, with the world fixated on the pandemic crisis, we cannot afford to put migration's long-term causes on the back burner. While the cost of responses may cause concerns, the cost of no decisions will certainly surpass that. There may be no clear, simple solution but having up-to-date evidence and data will surely help.

Safe water in danger zones

UNICEF is leading the way in highlighting and addressing the vital need for access to water and sanitation in conflict-affected regions

In recent years, humanitarian appeals have been increasingly dominated by stories and imagery of war: from Yemen and Syria, to longstanding and seldom-reported crises in Somalia or north-eastern Nigeria. The message has been increasingly clear: it is conflict, often combined with the challenges of a changing climate, that is at the centre of the major humanitarian crises of current times.

What is not always so obvious is how danger and suffering manifests itself for people on the ground. It is easy to focus on the immediate risks of bullets and bombs, but UNICEF research shows that in protracted conflicts, children under five are over 20 times more likely to die from diarrhoeal disease linked to unsafe water and sanitation than violence directly linked to conflict and war. Lack of adequate water, sanitation and hygiene (WASH) is a huge and under-recognised threat to the lives and long-term well-being of children in war zones.

"Globally, across fragile and conflict-affected contexts, 420 million children lack basic sanitation and 210 million children lack access to safe drinking water," says Kelly Ann Naylor, Associate Director for Water, Sanitation and Hygiene at UNICEF. "In fragile contexts, access to safe water and sanitation is often compromised; infrastructure is damaged, pipelines fall into disrepair and underdeveloped systems fail to meet immediate daily needs. Where no adequate water and sanitation services existed to begin with, the onset of conflict exacerbates the problem — particularly where water itself is a scarce resource, under increasing threat from a changing climate."

THE HUMANITARIAN CHALLENGE

Survival and development indicators differ starkly between children born into an extremely fragile context and those born into a stable, protected and developed country context. As of 2019, more than 800 million children live in 58 fragile contexts, including more than 220 million children living in 15 extremely fragile contexts. By 2030, it is estimated that more than 80 per cent of the world's poorest people could be living in fragile contexts.

"Humanitarian needs are on the rise: conflicts are more frequent, affect more people and last longer," says Naylor. "From a development perspective, fragile and conflict-affected contexts have the furthest to go. Children born in these contexts are more than eight times worse off across WASH indicators, including WASH-related indicators such as health, nutrition and education. Children living in these settings are also more than eight times as likely as children in non-fragile contexts to lack access to basic drinking water, and coverage of basic sanitation is even decreasing in nine fragile contexts. With these basic needs unmet, children fall ill, schools and hospitals cannot function, and disease and malnutrition spread."

One of the reasons WASH is so important is because it is about much more than water and toilets: lack of access to water and sanitation facilities contribute to a range of different humanitarian and long-term needs. Dirty water threatens health and is also a major factor in child malnutrition since diseases such as cholera prevent children from absorbing vital nutrients. A lack of safe female toilets in humanitarian settings puts women at risk of sexual assault and violence, and an absence of proper toilet and sanitation facilities in schools is a factor in low education rates for girls.

> *"In fragile contexts, access to safe water and sanitation is often compromised; infrastructure is damaged, pipelines fall into disrepair and underdeveloped systems fail"*

PREVIOUS PAGES AND ABOVE
Families queue for water in Somalia, just one example of a conflict zone where lives are threatened by a lack of access to water and sanitation

OPPOSITE
A child looks on at a water station in a camp for displaced people in northeastern Syria

The 2019 Ebola outbreak in the Democratic Republic of the Congo demonstrated how the spread of non-water-borne diseases can be exacerbated by lack of access to WASH. In a region affected by decades of civil war, and with severely inadequate WASH infrastructure, health centres without running water became hubs for transmission of the disease, and communities lacked washing and hygiene facilities to help protect themselves. WASH alone will not bring an end to these crises, but it has a crucial role in the humanitarian response: both in bringing the rates of transmission under control and helping prevent future outbreaks.

A THREE-PRONGED APPROACH
This consideration of both current and future needs is key. The growing number of protracted conflicts and recurring humanitarian crises has given new impetus to discussions of the humanitarian-development-peace nexus. The broad goal of triple nexus programming is to ensure that humanitarian, development and peacebuilding activities are brought closer together, blurring the traditional distinctions between the three, and acting in a way that tackles both immediate needs and longer-term challenges in fragile and conflict-affected regions.

It is recognised that WASH must adapt to support this approach. In a 2019 report, UNICEF identified key areas of focus, spanning from how WASH can help prevent conflict to how humanitarian and development WASH actors can ensure they reach the most vulnerable. Recommendations included the strengthening of ties between WASH and other sectors, for example, with child health during a cholera outbreak; the rehabilitation and maintenance of existing water and sanitation infrastructure rather than temporary alternatives; and continuing close engagement with communities in policies, planning and programming to ensure no-one is left behind.

Funding is, of course, an important consideration in all of this, and the amount of money allocated to WASH humanitarian programming in 2019 was less than 25 per cent of what is

currently needed. But the issue is not just how much funding is available, but how effectively WASH programming can adapt to play a key role in addressing the challenges of protracted conflicts. And this is a responsibility shared between a range of actors — including donors, national and local governments, humanitarian, and development agencies — to reduce the dangers faced by children in these settings, both now and for the long-term.

"We have been inspired by the volume of practical, implementable solutions that can be replicated and scaled up, from Bangladesh, Ethiopia, Lebanon, Nigeria, Somalia, the State of Palestine, Yemen and beyond," says Naylor. "And through these examples we can see the powerful potential of water and sanitation interventions that bridge the humanitarian– development divide and contribute to building peace. Our courageous colleagues and partners demonstrate that we can make this vital shift in what we do and how we do it, but only if governments, humanitarian and development partners, finance institutions, the private sector and communities find new ways to work together."

Empowered by design

UN Women has shown that smart infrastructure for women, designed by women, can make for a safer, more equal and inclusive society

The United Nations Entity for Gender Equality and the Empowerment of Women, also known as UN Women, is a global champion for women and girls, and was established to accelerate progress on meeting their needs worldwide. Achieving gender equality and empowering all women and girls is the fifth of the UN's 17 Sustainable Development Goals, but also integral to all dimensions of inclusive and sustainable development. All of the SDGs depend on the achievement of SDG 5.

Open, public spaces can be the heartbeat of communities. But in many cases, those spaces are unsafe for 50 per cent of the population, because of sexual harassment and violence. By including women in the designing of public spaces, we can ensure that their perspectives are embedded, and their needs are met. Through community efforts, UN Women is working around the world to help women claim their space.

"As female architects, we could make decisions that would lead to a better outcome, and address the needs of the community," says Dalia Osama, who worked with two other female architects to design a new community space in their Al-Shoka community of Gaza, Palestine. "For example, originally, people suggested putting the bathrooms on the left side of the garden entrance, but eventually, we decided to place them on the right, because the light was better from that direction, making the bathrooms safer for women and girls."

SUSTAINABLE SAFETY

Better infrastructure in parks, such as adequate lighting, clear sightlines and pedestrian pathways, as well as an open feeling to prevent entrapment, can go a long way towards improving the safety of park-goers. The architects – Dalia, Samah Al-Nahal and Nihal Zourob – worked together to create a blueprint of the public garden, with the help of 30 young people from the area, as part of a joint programme between UN Women and UN-Habitat, funded by the government of Belgium. The 2,600-square-metre garden in Al-Shoka opened to the public in March 2018, providing a space for Al-Shokans to enjoy the fresh air and feel safe and welcomed, in a space built by and for the whole community.

UN Women came about as part of the UN reform agenda, bringing together resources and mandates for greater impact. It merges and builds on the important work of four previously distinct parts of the UN system, which focused exclusively on gender equality and women's empowerment: the Division for the Advancement of Women (DAW); the International Research and Training Institute for the Advancement of Women (INSTRAW); the Office of the Special Adviser on Gender Issues and Advancement of Women (OSAGI); and the United Nations Development Fund for Women (UNIFEM).

In 2020, 10 years since the landmark General Assembly resolution that founded UN Women, and in the context of the global COVID-19 pandemic, the human rights of women and girls have greater prominence, universality and urgency than ever before.

Gender equality is not only a fundamental human right, but a necessary foundation for a peaceful, prosperous and sustainable world. The current global crisis has shown the many essential roles that women play, but also highlighted and heightened inequalities. "We must ensure that we use the lessons from past crises and do not simply rebuild the patriarchy," says Phumzile Mlambo-Ngcuka, Executive Director of UN Women. "We have the opportunity to address the pressing issues facing women and girls, and to chart a way forward that is more equal, more inclusive and more sustainable."

ABOVE
From left to right, Samah Al-Nahal, Dalia
Osama and Nihal Zourob, the three female
architects who designed the public garden
in Al-Shoka neighborhood in Gaza in
collaboration with community members

EDUCATION, RIGHTS AND EQUALITY **133**

ABOVE
The UNHCR's solar-powered
system is providing clean
water to Uganda's Bidibidi
refugee settlement

Sustainable sanctuary

A solar-powered water facility in Uganda, set up by UNHCR, shows how well-considered infrastructure investment can save the lives of refugees

When Asha Rose Sillah first arrived in Uganda as a refugee, there was so little water that she had to drink from a swamp to quench her thirst. Now, thanks to a new borehole project, her family has plenty of clean water to drink – and enough surplus to water the onions she grows for market. "I will have at least three to four bags of onions," she says. "There is more water in the community now, so we have time as women to do many things."

Asha fled South Sudan to Uganda's Bidibidi refugee settlement in Yumbe District in 2016 at the height of an emergency that saw thousands cross the border every day. Water was scarce then, making it hard for her to care for her five children. "There was a lot of illness," she explains. "We would drink any water we could find." Trucks delivered water from a source 100km away on poor roads, and refugees had to queue for hours to fill as many jerry cans as they could carry.

The minimum water requirement for a person is 20 litres per day. Three years ago, the supply in Bidibidi refugee settlement averaged just 2.3 litres a day. "From morning to noon, refugees could not cook, many didn't even have water to drink or take a bath," says Richard Ochaya, Senior Water and Sanitation and Health Associate for UNHCR, the UN Refugee Agency, in Bidibidi.

A PLENTIFUL WATER SUPPLY

Thankfully, things have changed. A solar-powered borehole, set up by UNHCR with investment from partners and the private sector, pumps groundwater to points closer to almost 500 households. "We have the capacity to pump 85,000 litres of water per hour, but we are only extracting 45,000 litres because we do not want to deplete the aquifers," says Ochaya. "We need to manage the resource and take care of the environment."

The plan is to one day hand over the facility to the Ugandan government to boost water supply in the district. "In future, if the refugee operations move away, the hosting community will be able to take care of such facilities and see them benefit generations to come," Ochaya adds.

A lack of adequate water, sanitation and hygiene facilities can devastate the health and survival of refugees in camps, outside of camps and in urban settings. The project in Yumbe District is an example of how smart investment can help refugees and host communities, by providing ready access to a supply of water.

The office of the United Nations High Commissioner for Refugees (UNHCR) was created in 1950, in the aftermath of the Second World War, to help millions of Europeans who had fled or lost their homes. Today, 70 years on – having won the Nobel Peace Prize in 1954 and 1981 – the organisation is still hard at work, protecting and assisting refugees around the world. Working out of three centres in Dubai, Copenhagen and Durban, it now has more than 17,000 personnel and operates in 135 countries, working to protect and assist refugees everywhere.

The UNHCR strives to ensure that everyone has the right to seek asylum and find safe refuge in another state, with the option to eventually return home, integrate or resettle. During times of displacement, it provides critical emergency assistance in the form of clean water, sanitation and healthcare, as well as shelter, blankets, household goods and sometimes food. It also arranges transport and assistance packages for people who return home, and income-generating projects for those who resettle.

A core part of UNHCR's protection mission is to guarantee access to adequate shelter in humanitarian emergencies. It provides tents, distributes plastic sheeting, and develops emergency strategies, tools and guidelines, rushing in emergency aid to those who need it most. On cold nights or hot days, its help can be the difference between life and death.

Constructing solid educational foundations

UNOPS is working to provide children with better learning facilities that are accessible to all

In April 2015, a 7.8 magnitude earthquake hit Nepal, taking more than 8,000 lives and damaging over a million houses. In 31 districts across the country lay the remains of 7,200 school buildings that had provided thousands of Nepalese children with a place to learn.

"We survived the disaster since it was a holiday. If it were other days, the scenario would have been dreadful," explained Hari, the principal of Saraswati Primary School in Nepal's Gorkha district. The further loss of life could have been immense – in 2008, an earthquake of similar magnitude killed more than 5,000 students and teachers in Sichuan, China.

Following the disaster, thousands of students in Nepal faced an interruption in their education while authorities decided on the best way to move beyond the destruction. As an interim solution, Temporary Learning Centres made from bamboo were erected so that children could continue their schooling.

However, the difficulties of studying in a Temporary Learning Centre are immense. Accessibility to the facilities is limited for children with special needs. Noise travels easily between the classrooms' thin barriers, disrupting lessons, and children sit on makeshift floors within walls that are not weather-proof.

Sustainable Development Goal 4 aims to ensure access to quality education and lifelong learning opportunities for all. This includes opportunities to acquire foundational and higher-order skills at all stages of education and development; greater and more equitable access to quality education at all levels, as well as technical and vocational education and training; and the knowledge, skills and values needed to function well and contribute to society.

To achieve this, school facilities themselves need to be of good quality, able to withstand shocks and stresses and provide a safe and sound learning environment. Moreover, facilities need to provide quality services – students' performance can be greatly enhanced by providing access to multiple infrastructure services (such as potable water, sanitation and electricity). Finally, providing access to basic services (for instance, safe transport) for the wider community can also support children's attendance at school. As a result, the provision of quality school buildings is just the first step.

MORE THAN BUILDINGS

At school, a lack of adequate sanitation facilities can lead to discrimination and a higher risk of gender-based violence. For instance, a lack of culturally appropriate water and sanitation infrastructure with safe, private, sex-segregated toilets, access to water and soap and facilities to change, wash or dispose of menstrual hygiene management materials may cause absenteeism amongst female teachers and students.

Access to ICT infrastructure at all levels of schooling fosters digital skill development, which is increasingly important for employment and entrepreneurship opportunities. Youth and adults who have relevant skills face better prospects in job markets that increasingly value technical or vocational proficiency. Internet access also provides teachers with a range of educational materials that can be downloaded and used to improve their skills and teaching methods. In addition, ICT infrastructure can also provide remote access to learning for those unable to travel far from home.

Providing access to infrastructure services in the wider community can further support students' attendance at school. For example, many young children, especially girls, are given domestic responsibilities like collecting drinking water or

RIGHT
The new schools were
designed to have low
operational costs and
to be fully accessible
to all students

firewood, leaving them no time to attend school. Modern and accessible energy and water infrastructure can save time for families and help ensure all children have the chance to pursue their education. Similarly, having to travel a long distance or along dangerous roads to reach school can expose women and girls to the risk of gender-based violence, making parents reluctant to send their daughters to school. Suitable transportation links can increase mobility by providing more options to reach school, even in locations farther away.

After the earthquake in Nepal, local authorities and communities were keen to not only rebuild the school facilities, but also build back better. With backing from government authorities and funding from World Vision International Nepal, DFID and the government of India, UNOPS worked in two districts to construct three earthquake-resistant, weather-proof schools that are fully accessible to all students, including those with special needs.

UNOPS implemented all work in collaboration with local authorities, developing and using local capacity throughout the project. For example, designs were finalised with a Kathmandu-based architectural firm and approved by the Ministry of Education and Ministry of Urban Development. The schools were also designed to have low operational and maintenance costs.

This work was carried out as part of a larger UNOPS response that includes assessing damaged buildings, reconstructing, retrofitting and modernising infrastructure, and providing socio-technical assistance. This is one of several school projects being undertaken by UNOPS.

Access to better facilities, with provision of quality reliable services, will give children the chance of a proper education in safe spaces, which will ultimately help secure their futures.

The building blocks of a secure, equitable society

Improvements to infrastructure can have radical effects on justice, equality and human rights, as UNOPS's work in Kosovo highlights

UNOPS has worked in Kosovo (referred to throughout this article in the context of United Nations Security Council resolution 1244) since 1999, helping partners achieve their development, peace and security goals. This includes the construction of a high-security prison between 2009–15, with a capacity for 300 detainees and built to international standards.

Targeted infrastructure interventions can contribute to reductions in violence, conflict and crime across societies. For example, a 2015 report by the SLOCAT (sustainable, low carbon transport) Partnership concluded that widespread electrification may enable steps to safety by deterring violence, while efficient roads and waterways allow law enforcement timely access to respond to emergencies and crimes. In regions where infrastructure may have been damaged by war or conflict, rebuilding infrastructure is a key step to restoring peace and order.

In Kosovo, the facility was designed to accommodate prisoners whose escape would represent a danger to the public or to national security. Sixteen buildings and work areas were constructed, including detention blocks, warehouses, medical and educational facilities, and rehabilitation workshops. The project was implemented above the minimal international standards for prison facilities, with maximisation of the size of cells, access to natural light, increased ventilation and welfare facilities.

Infrastructure should be designed to target particularly vulnerable segments of society; among the poorest, increased access can improve health and well-being by providing reliable basic services and allowing people to pursue livelihoods and economic opportunities.

Facilitating and modernising the provision of infrastructure services has been shown to reduce inequality. For instance, traditional gender roles may limit economic, social and educational opportunities for women and girls, who often bear the burden of household tasks like collecting water and firewood, cooking, cleaning and providing childcare. Accessible energy and water supply infrastructure can allow more time for the equitable pursuit of economic, social and leadership activities, and reduce time spent in unpaid domestic work.

BELOW
UNOPS supported the construction of a high-security prison in Kosovo, which has helped deliver peace and security goals

> *"Everyone, including the most vulnerable, needs to have access to institutions of governance to achieve participatory decision-making"*

Transportation also allows delivery of these services as well as the provision and administration of sexual and reproductive health supplies and services to all communities, including inaccessible or remote areas.

SUPPORTING EFFECTIVE INSTITUTIONS

Equally important to achieving peaceful and inclusive societies are strong and effective institutions. Constructing police stations, courts and prisons to adequate standards will improve law enforcement and access to justice. In Kosovo, UNOPS oversaw the construction of the prison facility, reviewing the prison's design completed by a local contractor, and provided procurement and project management support.

In addition, the project included the training of the facility's staff. For example, the local contractor was trained on planning and management of activities, as well as on quality, safety and environmental management techniques. UNOPS also offered guidance on maintenance during the prison's first year of operations to reduce the upkeep costs and ensure the prison facility's long-term sustainability.

The elimination of violence and armed conflict should be accompanied by strong institutional development at all levels of government, and universal access to justice, information and other fundamental freedoms. Everyone, including the most vulnerable, needs to have access to institutions of governance to achieve participatory decision-making. Such access may require better transport links – for example, to ensure all citizens can express their right to vote.

Access and accountability can also be improved with the help of digital communications infrastructure by allowing wider dissemination of information and online forums. Information and communication infrastructure may also enable marginalised groups to gain economic empowerment or greater influence in their communities. While digital technology can enable certain crimes – including cybercrime and illicit financial flows – it also provides the capabilities to combat them.

Eliminating discriminatory practices requires infrastructure related to governance and rule of law at the local and national levels to implement legal protections, enforce laws and ensure access to justice. Ensuring that everyone – including minority and vulnerable groups – has better access to paid employment, quality health services and real decision-making power in public and private spheres will further ensure that development is equitable and sustainable.

RIGHT AND OPPOSITE

The prison facility in Kosovo was designed to protect the public and national security

Making the connection

By sharing knowledge and pooling resources with multiple partners, UNOPS is providing hundreds of thousands of Iraqis with vital humanitarian information

In Iraq, more than 4 million people are in need of humanitarian assistance. Since January 2014, an estimated 3.3 million people have been forced to flee their homes to avoid conflict. For those affected, knowing where to turn for help is an ongoing challenge and the ability to stay connected is often a lifeline.

The Iraq Information Centre, implemented by UNOPS on behalf of global partners, addresses the needs of those affected by conflict. Using a toll-free number, people can get information on how to access humanitarian aid, including food, shelter, child protection and healthcare.

Since its launch in July 2015, the Iraq Information Centre has handled over 500,000 calls, connecting millions of people in Iraq with vital information and assistance. The centre handled more than 121,000 calls in 2019 alone, 24 per cent of which were from female callers. In addition, 47 per cent of these calls were from internally displaced persons, 30 per cent from returnees and 21 per cent from refugees. "Our goal is to empower communities in Iraq through easy and free access to information," says Muhammad Usman Akram, UNOPS Director in Amman. "The centre, which is one of the flagship projects of the Humanitarian Country Team in accountability, has connected over 3 million people in Iraq with the humanitarian agencies within the past five years."

In 2015, for example, the call centre was contacted by people stranded in Sinjar in northern Iraq. Hundreds were caught in fighting between the Islamic State of Iraq and the Levant (ISIL) and Peshmerga forces from Iraqi Kurdistan. Details from the calls were shared with humanitarian aid agencies in the country. Through UN negotiations, 400 people were granted safe passage from Sinjar to the Kurdistan Region of Iraq.

The centre has also been integral to COVID-19 response activities. In collaboration with various humanitarian clusters, it has supported SMS campaigns to deliver key information on

UNOPS is working with
global partners to ensure
that displaced Iraqis can
access essential advice

OPPOSITE AND BELOW

The Iraq Information
Centre is one of the
biggest UN inter-agency
initiatives of its kind

prevention measures across Iraq. Moreover, the centre has served as a voice to thousands of people impacted by the COVID-19 pandemic by collecting feedback from callers on response efforts and presenting it via a COVID-19 dashboard. To date, the dashboard has logged more than 8,000 calls related to COVID-19, with the majority of calls related to affected livelihoods and loss of employment opportunities. This gives humanitarian actors a better understanding of the challenges faced by the population.

"Beyond listening, accountability within the humanitarian sphere is about ensuring that the humanitarian response is relevant to the needs of the affected population," says Bana Kaloti, UNOPS Middle East Regional Director. "Together with our partners, UNOPS is proud to be able to contribute to the more efficient and effective delivery of humanitarian assistance in Iraq."

The information centre is one of the largest UN inter-agency projects of its kind, helping to get life-saving information to those who need it most. Through a similar sharing of knowledge, technologies and technical skills, countries can bring actors together

and capitalise on global partnerships to ensure the most effective implementation of sustainable development policies.

Achieving the 2030 Agenda targets will not be possible without mobilising all available resources – this includes a renewed global partnership that brings together governments, civil society, the private sector, the UN system and other actors.

A RENEWED GLOBAL PARTNERSHIP

Strengthening partnerships at the national or international scale requires infrastructure that can facilitate integration through information sharing, knowledge transfer and capacity building, incorporating a variety of infrastructure types. The targets for SDG 17 (partnerships for the goals) can especially be supported by transportation and digital communications networks.

In Iraq, UNOPS has implemented digital communications infrastructure in collaboration with dozens of UN organisations, non-governmental organisations, government authorities and three telephone companies to provide the most up-to-date information and ensure a reliable mobile service across Iraq.

Outputs of the project include not just the information centre itself, but also a database of information to support humanitarians, the establishment of knowledge sharing networks between humanitarian aid agencies, and the establishment of coordination mechanisms between the government and telephone companies.

In addition, to address sensitive topics, the call-centre team has received training on the prevention of sexual abuse and exploitation, gender sensitivity, mine-risk education and child protection. The information received from the calls is shared with relevant UN organisations and non-governmental entities to address service gaps and ensure a coordinated response.

As a result, the project is advancing the achievement of SDG 17 targets, as well as several targets set by other SDGs, including SDG 1 (no poverty), SDG 3 (good health and well-being) and SDG 5 (gender equality). Beyond creating partnerships and dialogue, the project also contributes to the reduction of poverty, hunger and inequalities, and promotes good health, peace, justice and strong institutions – illustrating how the impact of sustainable infrastructure extends past individual sectors to enable development.

Strength in breadth

Technical knowledge will always be essential, but social and environmental considerations are playing an ever-more significant part in the education of engineers

SDGs: 4, 9, 11

In response to global challenges such as climate change, population growth and urbanisation, the engineering profession is constantly developing in multiple (often competing) ways; for example, in technological advancement, environmental responsibility and social responsibility. While the challenges are not new to civil engineering, advances in technology and changing societal views on acceptable solutions are providing different ways to respond.

Mainstream awareness of the detrimental impacts of the industrial age gained traction in the 1960s and '70s (with new environmental legislation, such as the US National Environmental Policy Act of 1969), and has been growing since. Initially, however, there was not widespread political support or the public debate necessary to fundamentally change the course of development decisions in the western world. "Sustainable development" had not yet been defined in the way it is broadly understood and accepted today.

The introduction of the Sustainable Development Goals (SDGs) in 2015 was the result of decades of learning and global diplomacy. The 17 SDGs provide a framework on which to measure the merits of development decisions and outcomes. Economic reasoning still dominates many development decisions, but it is now more common to find social and environmental considerations (reflected in the SDGs) influencing infrastructure outcomes. However, the SDGs are broad-reaching in their ambition and complex to implement.

"Since the SDGs were introduced, there has been greater recognition of the need to teach a broader range of skills to our engineering students," says Dr Kristen MacAskill, Lecturer in Engineering, Environment and Sustainable Development at the University of Cambridge's Department of Engineering. "They need to now understand their place in society, and not work in silos, as used to be the case."

TRAINING BEYOND THE TRADITIONAL
To respond to this complexity, there is a need to train civil engineers to engage with complexity beyond traditional engineering. The American engineer Samuel Florman explored the limitations of traditional engineering training in the 1970s, suggesting that education in the arts and humanities would help address the "naivety" of the engineer and force the consideration of "the imperfections and absurdities of his fellow human beings". The underlying point Florman makes (somewhat provocatively) is that developing the competence to acknowledge and engage with social context is critical in an engineer's training.

Engineering schools are now attempting to prepare students not just for technical competency, but to consider the wider implications of their decisions. Engineering curricula today include modules such as "The Engineer in Society", "Climate Change Mitigation" and "Energy and the Environment". There is disagreement about how to structure engineering education, with an apparent trade-off between how much to emphasise traditional technical knowledge versus newer sustainability considerations. Many will argue that further change is needed to support sustainable development concepts in engineering education; others will counter that the technical focus of an undergraduate engineering degree must not be jeopardised. This may never be completely resolved, but healthy debate is useful to work out how to address these needs together.

Further learning through postgraduate education has a key role to play. Professional practice courses – that assume participants have a basic level of technical competency through prior undergraduate training – can help develop the broader skills needed.

"We are training professional engineers to engage with broader societal issues," says Dr Kristen MacAskill. "We've found they increase their contextual awareness of the role that engineering plays in society, and use this to explore leadership."

While technical competency remains core to civil engineering, members of the profession need to have the capacity to step out to wider roles to inform political and public agendas – to allow the profession to be heard more. This helps support not just the practical application of science, but also to help achieve the objectives of the SDGs.

ABOVE
Engineering is facing complex
challenges that are putting
traditional approaches to
training in question

For infrastructure such as a communal water point to be effective it must take the needs of its end users into account

Sustainable engineering for all

If engineering is to meet a society's needs, it must understand and engage with all aspects of that society

SDG: 5

Sustainability is often described as having three pillars: economic, social and environmental. Civil engineering has been defined by the great railroad constructor Thomas Tredgold as "working with the great sources of power in nature for the use and convenience of society". Both descriptions address the fact that civil engineers need to consider the needs of society – but society is not uniform. In the past, engineers may have been good at distinguishing between type of rocks, but perhaps less adept at recognising the differing elements of the population. This is now changing.

Understanding these differing elements and their needs is critical in the provision of basic water and sanitation facilities in low-income countries. Women traditionally are responsible for domestic work, which includes collecting water and dealing with the sanitary requirements of children and infirm relatives. Not recognising their specific needs can lead to project failure. Wasted resources can result if projects are not thought through. For example, communal water points may be located in the wrong place, leaving women open to harassment and rape; toilets may be provided that cannot be cleaned due to bad design; or busy people may be expected to volunteer to run public facilities.

Better outcomes come from involving women, and regarding them as clients. Research by Loughborough University describes how water and sanitation impact on the rest of women's lives: the effort and time taken in collecting water; the impact of unsafe water on their health; and the reduction in educational opportunities if there are no suitable toilets for girls at school.

THE NEED FOR CONSULTATION

Despite recognising the important role of women at policy level, putting this into practice has taken time to become standard engineering practice. There may be courses in engineering mathematics, but engineering sociology is not widespread.

Working with social scientists is a good start, but they lack training in engineering issues, so often only offer socio-economic solutions, such as having women on management committees or consultation with women's groups. This may involve women in the process, but does not necessarily change the design or construction of the facilities.

A change of approach came when attention was turned to another socially excluded group. People with disabilities in low-income countries need to be able to access water and sanitation facilities. While there are social aspects to this, it is a mainly physical challenge. The location, design and type of infrastructure had to be adapted and, to understand the problem, the users had to be directly consulted. This meant civil engineers had to be involved in the discussions and standard designs had to change.

This change in role for engineers has also influenced the design of facilities for women. One physical – or rather biological – difference between men and women is menstruation. There are many social aspects to menstruation: women need safe spaces to manage their menstrual hygiene; they need water for laundry and bathing; and they need discreet waste-disposal methods. This has major impacts on girls' willingness to go to school during their period. These are topics that engineers can and have engaged with, being practical challenges requiring innovative solutions.

Sewer blockages can also have a gender dimension, as they can be caused by flushing wet wipes, sanitary products or (with women still taking on the bulk of cooking in many cultures) by pouring cooking fats, oils and grease down the drain. Everybody needs to be engaged in preventing this problem.

SDG 5 aims to "achieve gender equality and empower all women and girls". Engineers can contribute towards this through the design and construction of accessible infrastructure, easing the burden of collecting water, improving access to education and providing suitable waste disposal.

The pros of cons

Tideway, the company delivering London's "super sewer", is committed to employing people with convictions – an initiative that will benefit the entire country

SDGs: 8, 17

The Thames Tideway Tunnel is a major new sewer designed to protect the River Thames from the millions of tonnes of sewage that currently spill into its tidal section every year. Tideway is the new company that has been established to build the tunnel. As part of the delivery of the tunnel, a comprehensive legacy programme has been drawn up that sets out to deliver sustainable benefits that reach far beyond just cleaning up the river. These include employing those from vulnerable groups such as people with criminal convictions.

According to the National Police Database, more than 11 million people in the UK have a criminal record. This is around 20 per cent of the working age population (those aged 16–64). One in three men and one in nine women have a criminal offence by the age of 56. People with convictions make up a sizeable proportion of the unemployed population and are often overlooked by employers. In the past ten years, 33 per cent of Job Seekers Allowance claimants have a criminal record.

Tideway is committed to removing the barriers to employment that are often faced by people with convictions and recruiting them into sustainable jobs (those lasting six months or more). It matches people with convictions to suitable jobs that help deliver the project and achieve social value. Tideway has a target of employing one person with convictions for every 100 employees.

STRATEGIES FOR SUCCESS

Tideway employs and advocates various strategies found to be effective in helping those with convictions, including Release on Temporary License and Ban the Box, which removes the requirement to declare spent convictions on an employment application. Tideway and its suppliers have partnered with several charities, such as Changing Paths, Bounce Back, Key4Life and

Nacro, to help deliver this challenging target and establish the processes to support people through this employment route.

The company encourages the employment of people with convictions in more than one way. Its staff have volunteered their time to attend workshops in prisons with the intention of growing the confidence and employability of inmates. These workshops also help to provide business skills to prisoners due for release and challenge industry perceptions around employment of this group.

To date, the project has supported 34 people with convictions into sustained employment, several of whom have gone on to other jobs, with a positive Social Return on Investment estimated at £6.86 per £1 spent. Tideway has mapped its legacy programme against the UN Sustainable Development Goals (SDGs), and through this work with vulnerable groups, including people with convictions, it is directly contributing to SDG 8 – decent work and economic growth.

Re-offending is estimated to cost the UK around £18 billion every year and programmes such as this help people into employment and away from re-offending. As one of those employed by Tideway through this route said: "I want to be here while my kid is growing up, I want to set a good example to my family."

This employment strategy garnered external recognition when it received the Employer's Forum for Reducing Re-offending Award for Business Working with Local Communities. It was also highly commended in the Corbett Network awards.

Tideway has realised extensive social benefits and helped achieve SDG 8 by working innovatively to help people achieve their best by working on an infrastructure project that is going to protect the River Thames from sewage pollution. This should ensure a long-lasting legacy for both participants and the environment for many years to come.

OPPOSITE
Tideway operates several policies that encourage the employment of those with convictions

A catalyst for change

The Africa Catalyst project is addressing the region's engineering and infrastructure challenges through partnership, training and development

SDGs: 8, 9, 10, 17

In 2018, 10 million people left their homes in sub-Saharan Africa, a region that hosts over 18 million refugees – more than 26 per cent of the world's refugee population. That number has increased constantly in recent years due to unresolved and cyclical conflicts across the region. This creates a huge humanitarian demand for engineering solutions, to address basic needs.

"The escalating climate emergency and the impact of rapid urbanisation has been increasingly felt," says Jo de Serrano, CEO of the disaster relief NGO RedR UK. "Civilians are always the first responders when disasters strike. Engineers and engineering institutes exist in huge numbers across Africa, but in a humanitarian setting they are under-utilised. They need to be there before emergencies strike, as well as at the front line of response to humanitarian crises in areas such as water, sanitation and shelter. We have been working tirelessly on bringing together the humanitarian and engineering sectors for 40 years."

The Africa Catalyst project is a RedR UK capacity-building project designed to strengthen the Federation of African Engineering Organisations (FAEO) – the umbrella body for 33 professional engineering institutes (PEIs) across Africa. Funded by the Royal Academy of Engineering, supported by the Global Challenges Research Fund, the project aims to improve engineering across sub-Saharan Africa, especially for humanitarian responses to disasters.

"The Africa Catalyst project has reignited the FAEO's vision of building its capacity in engineering to meet the requirements for achieving the 'Africa we want'," says Martin Manuhwa, FAEO President, "an Africa with adequate, safe and resilient infrastructure, zero hunger, healthy citizens, zero tolerance to corruption and endowed with competent human capital, especially in engineering, to meet most of the targets of the Sustainable Development Goals by 2030 and the African Union Agenda 2063."

Good engineering work in turn enables sustainable economies to develop. "There is also a strong positive correlation between engineering strength and economic development," says de Serrano. "Strengthening engineering bodies gives scope for potential

RIGHT
RedR UK aims to continue to support
the engineering profession in Africa

growth, drives economic and social development, facilitates education and healthcare, and enhances quality of life."

TRAINING AND DEVELOPMENT

Taking a collaborative approach, RedR UK has been working with the FAEO to strengthen its financial, fundraising, governance and diversity capacity. Initially, the FAEO's needs, capacity gaps and key areas were assessed to prioritise training and development. Throughout 2019, FAEO then developed key organisational policies, procedures and knowledge with RedR UK expert trainers.

Katie Bitten, RedR UK Programme Coordinator, explains the organisation's learning approach. "Much of the learning has been in real-time," she says. "For example, we used an opportunity to apply for funding as a way to coach FAEO staff on proposal writing and budgeting. We've found this to be an effective approach, contextualising the learning and giving it greater impact while also ensuring it is FAEO-driven, rather than a top-down capacity-building approach. It's been fantastic to work collaboratively with the FAEO and see the huge difference in the organisation from two years ago."

The FAEO is now being helped to deliver learning events and programmes to engineers across Africa – most recently, the team delivered a workshop to young engineers, responding to the challenge of getting enough people into the profession.

The FAEO's Rotimi Famisa, Business Development Manager, and Rose Nungul, Marketing and Membership Affairs Manager, were thrilled by the success of this event. "Participants rated the event as excellent," says Famisa. Nungul adds: "This is very encouraging as we are poised to deliver future trainings on topics such as innovation and technology in engineering, the importance of engineering diversity, inclusion

and accessibility, and sustainability in engineering, for professional engineering institutions in each of the five regions of the FAEO."

Jo de Serrano is focussed on the development of more programmes of this kind. "RedR UK will continue to work with the FAEO to strengthen its capacity over the next year," she says. "We also hope to do similar projects with other engineering institutes in Africa, as well as integrate more closely the humanitarian, development and peacebuilding sectors with the engineering profession in Africa."

Leading the way in pooled knowledge

UNIVERSITY OF LEEDS
Location: Leeds, UK
www.water.leeds.ac.uk

Ten years ago the University of Leeds was looking at some of the global challenges that it could put its considerable weight and knowledge towards trying to understand and tackle. One of those chosen issues was water, and it created water@leeds, an interdisciplinary body that draws on knowledge from across the university's different faculties and directs it towards the challenges of the world beyond.

Since it was founded with £1 million of seed money from the university, water@leeds has gone on to generate more than £12 million of funding per annum and provide research for between five and six thousand companies across the globe. It also has around 180 students doing doctoral-level research.

"We operate as an umbrella across all disciplines and provide an interdisciplinary offering in research to the outside world," says Professor Martin Tillotson, Chair in Water Management and Director of water@leeds. "I came to the role with 10 years' experience in the water industry, where I found it frustrating to engage with academia. I knew the expertise was out there but finding it was challenging. We are now a single point of contact and we can offer companies different

experts from across campus to address their concerns and interests. They don't have to work with different groups of economists, engineers and scientists – we can do it all, saving time and money."

Early successes saw the centre start to develop credibility and build a strong reputation, both inside the university and among external funding bodies and companies. water@leeds bases its work around four missions – research, impact, next generation and partnerships. Impact refers to the non-academic benefits of the research, looking at the research in terms of job creation, income generation and environmental benefits. The research also covers the entire water cycle, from climate and meteorology, through to water catchment, agriculture, infrastructure and the built environment, rivers and lakes, water supply, sewage and outfall.

"It's the whole cycle, which is fairly unusual for academia worldwide," says Professor Tillotson. "We have built momentum over the past 10 years and it's turned us into the centre we are today. Our current research portfolio is worth around £70 million of live grants. That makes us the largest in the UK and one of the largest in the world."

Applied academia

GLOBAL ENGINEERING DEANS COUNCIL
Location: Worldwide
www.gedcouncil.org

On 9 May, 2008, 27 academic leaders of engineering schools and colleges, from diverse cultures and national origins, inaugurated the Global Engineering Deans Council (GEDC) by signing the Paris Declaration. The document acknowledges transformations of planetary magnitude caused by the exponential demographic growth and planet-wide means of collective action, enabled by increasing technological effectiveness. It also observes that these engineering achievements pose engineering challenges, such as food supply, clean water, accessible healthcare, security, cleaner energy, cleaner environment, changing demographics, climate change, and sustainable development; essentially, foreshadowing the list of UN Sustainable Development Goals.

"The foundation of GEDC is thus the motivation to provide a global platform for academic leaders of the engineering field to address the SDGs primarily by nurturing the development of adaptive engineering leaders," says Professor Şirin Tekinay, Dean

OPPOSITE

Dori Tunstall, Dean of the
Faculty of Design at OCAD
University, delivers a
speech to the GEDC, 2019

ABOVE

The panel for "Women in
Engineering and STEM"
at the GEDC Annual
Conference, 2019

of the College of Engineering at the American University of Sharjah, and Chair of the Global Engineering Deans Council.

What started with 27 visionary founders today spans the globe, with about 500 members from all continents; roughly 50 countries. The council brings together global academic leaders of engineering (deans, university vice presidents, centre directors, and leaders of similar stature) with captains of the industry; decision makers of global employers of engineers and manufacturers of products for engineering education, research and development, consultants in innovation, entrepreneurship, and career development.

The GEDC annual conferences – co-located with World Engineering Education Forum (WEEF) by the International Federation of Engineering Education Societies (IFEES) – have continued to be exciting summits for GEDC members over the last 12 years. These conferences are held in different parts of the world, for instance, in New Mexico, USA in 2018, Santiago, Chile in 2019, and coincidentally, during the 75th birthday of the UN, in Cape Town, South Africa. The geographic diversity is united in thematic focus on different aspects of the impact of engineering education, research and development, innovation and design, on a global scale. The 10th anniversary conference in 2018 in New Mexico was themed "Peace Engineering", which is a theme that has garnered worldwide traction and increasing support.

Peace Engineering has in heart the planet's sustainable future: it is, essentially, the achievement of the Sustainable Development Goals by applying science and engineering principles and thus generating engineering solutions, technological tools, to transform and improve life on the planet, and social well-being of humans.

GEDC was founded on, and operates with, a sense of basic responsibility that it is the mission of all engineering education institutions to raise the new generation of global leaders, in as much orchestration with industrial and government sectors as possible. The GEDC has provided the fertile ground for focus groups such as the Carbon Free Innovation Network that drew participation from North America, Asia Pacific, and Europe. Most recently, the global engineering community has been excitedly unified in the vision statement issued by IFEES. The definition and pathways of Peace Engineering are given best in this recent "IFEES Peace Engineering and Sustainability Declaration".

"The GEDC looks into the future of mounting challenges our planet faces," says Professor Tekinay, "and continues to serve as a conduit for engineering leaders to be empowered in their respective regions, to support each other globally, to deliver education at global quality standards, and to collectively generate socio-economic and environmental impact."

Hands-on
education

KINGSTON UNIVERSITY
Location: Kingston, London, UK
www.kingston.ac.uk

Ensuring that students "belong", gain a broad knowledge through hands-on experience and become employment-ready is a top priority at Kingston University's Faculty of Science, Engineering and Computing. "We have many first-generation university students and one of the challenges is inspiring them with confidence," says Faculty Dean Dr David Mackintosh. "We have a strong track record of developing individuals irrespective of their background. There's no doubt about their ability – but they may not have been told before that they can succeed."

To this end, mentoring the yearly intake of around 100 or so undergraduates (full time and degree apprenticeship students) is vital. Recent graduates are invited to talk to new students about the importance of perseverance – especially when things get tough. "They know what the course involves," says Dr Mackintosh, "so they can be honest about how they overcome challenges during the course and grow in confidence. We get new students engaged from the start, to help them feel they belong. Teamwork is imperative, too, rather than them getting lost in a large year group."

Based in south west London, Kingston has a long history of teaching civil engineering. In 1942, Kingston University's predecessor, Kingston Technical College, introduced a full-time diploma enabling servicemen returning from the war to obtain academic qualifications. It also laid the foundations for the degrees of today's School of Engineering and the Environment.

The focus is still firmly on engineering graduates gaining a broad knowledge through hands-on experience of real-world problems underpinned by a theoretical understanding of their subject. Employers are directly involved in shaping the curriculum to ensure that it remains relevant. As students finish their final year, they are assessed on their project work by a panel of experts from top industry firms.

Apprenticeships are a crucial part of – and help to strengthen – a curriculum that is taught by professionally qualified academic staff. A number of Kingston students were involved in the development of Heathrow Terminal 5 and also the new Wembley Stadium. Professor George Haritos, Head of the School of Engineering and the Environment, describes this active involvement as invaluable experience that benefits both students and employers. "They make tangible contributions and increase their chance of a good job at the end," he says. "It's like an extended interview."

Kingston ensures that students cover a broad range of subjects, including geotechnics, water engineering and sustainability, demonstrating its emphasis on obtaining practical experience. For a project on a brownfield site, for example, the undergraduates will examine ground conditions first. Then they will use their findings to determine the "skeleton" of the construction suitable for the site and the best materials to produce it.

"Everything they do has to be commercially viable and practical – and follow green targets," says Professor Haritos. "It's about looking at available resources nearby to minimise the impact on the environment while ensuring that the construction is serviceable and user-friendly. There's no point in a building getting cold in winter and hot in summer."

Just as every new building requires solid foundations, the next generation of engineers need a solid practical and academic base. Whatever their background, those graduating from Kingston University are assured of a future career that is grounded in success.

An elite engineering university

POLYTECHNIQUE MONTRÉAL TECHNOLOGICAL UNIVERSITY
Location: Montréal, Canada
www.polymtl.ca

Polytechnique Montréal prepares for its 150th anniversary in 2023, with sustainable engineering an increasingly important part of its offering. The creation in 2019 of the Institute of Sustainable Engineering and Circular Economy rests on more than three decades of exploring the subject. "The circular economy targets the closure of loops in the treatment and disposal of materials," explains Philippe Tanguy, President of Polytechnique Montréal. "The de-globalisation of various production chains means that sustainable engineering is bound to grow and become our priority."

With nearly 300 faculty teaching across 120 programmes to 6,800 undergraduates and 2,300 postgrads, this elite Canadian engineering university targets specific applications of sustainable engineering. Staff have driven research in a wide range of topics, including the capture and conversion of carbon dioxide from industrial sites; the development of technologies and processes in water treatment; the recovery of rare-earth metals from electronic consumer waste; the use of hydrogen in green chemistry; the electrification of sustainable transport; the management of urban waste; and the development of printed organic electronics. When Canadian astronaut and alumnus David St-Jacques joined the International Space Station in 2018 (pictured, right), he was proud to display his Polytechnique Montréal pin badge.

The university has 12 Industrial Chairs from Canada's National Research Science and Engineering Council, 24 Canada Research Chairs, one Canada Research Excellence Chair and 10 private Industrial and Philanthropic Chairs, making it one of the top three engineering schools in Canada and Quebec's leading university for R&D in engineering. Its TransMedTech Institute provides a living lab and integrated environment to support interdisciplinary collaborative processes for the co-creation, development and validation of new medical technologies, while the Institute for Data Valorisation (IVADO) brings together over 1,000 scientists from industry and academia to develop cutting-edge expertise in data science, operational research and artificial intelligence.

A third of students are from overseas and the university is actively promoting student mobility. It is also involved in major international training programmes and R&D partnerships with industry and other top-ranking universities. "We are keen to foster greater connections with industry and the international research community to join forces in addressing universal issues," says Tanguy.

Full STEM ahead

ST JAMES SENIOR GIRLS' SCHOOL
Location: London, UK
www.stjamesgirls.co.uk

It has been estimated that the under-representation of women in STEM subjects – science, technology, engineering, maths – costs the British economy at least £2 billion in lost labour. This is in addition to denying the UK's science-based industry the brains, inspiration and drive of a significant section of the population.

One school working to change this is St James Senior Girls' School in Kensington, London, an independent school with just over 260 pupils, which encourages girls to explore STEM subjects as one part of a suite of opportunities that significantly contribute to the girls' positive development and their future prospects.

"The school achieves excellent results but it is also very strong on pastoral care and support, and in encouraging the girls to have a go," says Tomas White, Assistant Head (Co-curricular) and Head of Chemistry. "At mixed schools, girls sometimes hold back as they worry about making a mistake but we encourage them to think they can try anything. There is no stigma to having a go and getting it wrong as that is part of the learning journey."

As well as providing a broad curriculum, sporting opportunities and the Duke of Edinburgh's Award, the school's offering includes the Minerva Society, which features regular talks by external speakers, who are often female role models. One speaker was a former pupil who was studying maths at Imperial College. Having been educated in the inclusive, stimulating environment of St James, she didn't realise that maths was considered by some to be a "male" subject until she arrived at university.

St James also has several annual days for STEM activities, during which girls are put into teams and asked to work together to solve engineering or programming challenges. The onus on collaboration is a core element of the school's philosophy, with pupils encouraged to work with one another rather than to compete against each other. The school's STEM club is another popular activity, with girls using a range of tools and materials to take on long-term projects.

As Kenneth MacLean, Head of Science and Biology, explains, "Crucial skills in engineering include communication and working as a group. We encourage the same kinds of approach that an engineer might take."

A collaborative school of thought

LASSONDE SCHOOL OF ENGINEERING, YORK UNIVERSITY
Location: Toronto, Canada
www.lassonde.yorku.ca

"Renaissance engineer" – that's the term that is advanced at York University's Lassonde School of Engineering in Toronto as its aspiration for the next generation of engineering students.

For Dr Jane Goodyer, the school's Dean, a renaissance engineer is one equipped to tackle the challenges facing the world through exposure to a wide range of educational experiences; one who seeks out people with different knowledge or ways of thinking. "We are open and creative," she explains. "We have passion and we have perspective. Bring that together and you can solve complex global issues. You can't solve these things through one technical lens, you have to be interdisciplinary and that's something we are developing in our next generation of engineers."

One member of this new generation is Maira, an international student from Dubai, who was impressed by the range of experiences offered by the Lassonde School of Engineering. "The university appealed to my passions," she says. "I care about science and technology because I care about people; I also care about literature because I care about people. When I researched the work being

done at Lassonde, I saw that it recognised the problems facing the world and it had the chutzpah to deal with them. People here seem dissatisfied with the status quo and want to make meaningful change in the world."

York University is highly engaged with many of Toronto's local communities and its commitment to improving the quality of people's lives has made it a leader in social responsibility and sustainability. The university was rated 26th in the world and 5th in Canada in the *Times Higher Education*'s new Impact Ratings for 2019, which measure higher education institutions against the UN's Sustainable Development Goals.

"These ratings highlight how the global higher education sector is contributing to international efforts to build a more socially, economically and environmentally sustainable world," says Dr Goodyer. "We were established in 2011, so we're a relatively new engineering school, and within that foundation is a commitment to create change for our students and the world through our passion and the principles of social justice. That is unique and you can feel it round the university."

The school's "renaissance" approach is very much part of this university-wide spirit; a spirit that directly informs the collaborative nature of some of its most innovative courses. York's capstone programme for final-year students, for example, has spawned its C4 (cross-campus capstone classroom) learning environment. This enables participants to work with senior students from other faculties to tackle specific challenges as part of an interdisciplinary team.

Four of the current C4 projects include students from the Lassonde School of Engineering: an inflatable solar energy collector; a community ride-share system; an anti-matter containment unit; and a wastewater recycling system. As Dr Goodyer explains, each project team is made up of students from a host of disciplines working together and sharing their distinct perspectives. "When we work on waste water," she says, "we bring together environmental, arts and engineering students to find ways to use waste water and how to communicate the problems."

It's a clear illustration of the school's hugely successful "renaissance" approach. An approach that is better preparing the engineers of tomorrow for the complex challenges that lie ahead.

Engineering change amid the climate emergency

SYRACUSE UNIVERSITY COLLEGE OF ENGINEERING & COMPUTER SCIENCE
Location: Syracuse, New York, USA
www.eng-cs.syr.edu

Across the world, today's youth protests against the climate emergency are drawing attention to some of tomorrow's brightest minds. Syracuse University's College of Engineering and Computer Science aims to focus such minds on tackling the challenges we face moving forward, at a local and global level.

"Environmental challenges are among the world's most daunting challenges in 2020," says J Cole Smith, Dean of the College of Engineering and Computer Science. "Climate change threatens to affect all facets of our profile of natural resources, including freshwater availability and distribution. These challenges promise to become increasingly acute in the coming decade. Nations that do not address freshwater management risk problems ranging from an inability to support water and food resources for its people to international conflict over scarce water access."

In 2018, the university launched its own Infrastructure Institute to support the provision of modern and socially responsible public infrastructure across the USA. One of its leading researchers is Shobha Bhatia, Meredith Professor in Civil and Environmental Engineering. She was dubbed a "Geolegend" in a recent issue of *Geostrata*, the magazine produced by the Geo-Institute of the American Society of Civil Engineers (ASCE). An advocate of engineering education, she is also co-director of the university's Women in Science and Engineering (WiSE) initiative.

Professor Bhatia is committed to investigating new and sustainable ways to preserve water. Her research addresses problems of soil erosion and develops techniques to filter and decontaminate water reserves in developed and undeveloped countries. But rethinking infrastructure is an upstream struggle. "I think it takes time for industry to see the values," she says, "and it takes a long time

for people to be convinced. You have to be persistent; but I'm going to be here for a long time."

Syracuse students enjoy the best of both worlds in an urban location close to the Adirondack mountains, the Great Lakes and the Finger Lakes. "This is a uniquely terrific place to engage in civil and environmental studies," says Cole Smith, "with the challenges and benefits of having four full seasons."

The perfect setting, then, for young engineers to help tackle the climate emergency.

Guarantors of sustainability

SENECA COLLEGE
Location: Toronto, Canada
www.senecacollege.ca

Whether you are a regular citizen, technical practitioner, decision-maker or someone with the highest professional expertise, everyone has a role to play in sustainability. These are practices that benefit society, infrastructure upgrading, personal healthcare and more.

Seneca is a Canadian postsecondary institution that annually trains and educates more than 30,000 full-time students from more than 150 countries. It offers more than 500 career options across multiple campuses in the Greater Toronto Area. "Seneca's portfolio of science and technology programmes have sustainability embedded in their design," says Ranjan Bhattacharya, Dean of the Faculty of Applied Science and Engineering Technology (FASET). "Graduates are well positioned to lead in industry long-term sustainability initiatives."

Graduates from Seneca's diverse technology programmes in FASET are examples of what has been described as "life-cycle guarantors of sustainability". For Andrew Wickham, Green Citizen Project Manager in FASET: "These include building standards based on LEED criteria, automated systems, transportation, water and wastewater systems, remediation of contaminated sites and low-impact development technologies."

With expertise ranging from civil and electronics engineering to building systems and environmental practices, they ensure that work meets the sustainability performance specifications for long-term use. In addition, FASET develops and promotes sustainability projects that provide intrinsic and educational value. As Seneca strives to support and promote the UN's Sustainable Development Goals, students are introduced to sustainability in many ways. "One example is the Green Citizen initiative from FASET, which embeds an environmental imperative within our academic programmes," says Wickham. "Our annual Green Citizen speaker and workshop symposium recently looked at the necessity for resilience in the design and maintenance of both built and natural environments. It also celebrated how air, water and soil are part of a shared commonwealth of healthy environmental attributes supporting social enhancement goals."

This and many other sustainable initiatives on Seneca's campuses – including beehives and honey, environmental landscape planting and an aquaponics urban farming unit – will continue to position Seneca as a leader in this wide field that pushes the envelope to foster a brighter future.

Educating agents of global progress

INSTITUTO SUPERIOR DE ENGENHARIA DO PORTO (ISEP)
Location: Porto, Portugal
www.isep.ipp.pt

There are few issues that motivate and inspire young people as much as sustainability. That's why the subject has become a core focus of Portugal's Instituto Superior de Engenharia do Porto (ISEP), one of Europe's leading engineering schools with more 6,000 students.

"Sustainability is very important to our students," explains Professor Maria João Viamonte (pictured, right), Dean at ISEP since 2018, and the first woman elected to that role. "It is important that we are concerned with sustainability now so we can have a better future for them as well as ourselves. As future engineers, we believe our students can produce creative solutions for present or upcoming challenges, becoming agents of global progress."

All of the degrees offered by ISEP have one or two courses that cover topics related to sustainability, often directly informed the UN's SDGs, and there are two Masters degrees that focus specifically on sustainability – one on sustainable energies and the other on development practice. On campus, ISEP has implemented a strong internal campaign to reduce plastic, water consumption and waste, while research groups at ISEP are involved in several sustainability projects such as REWATER and BIOFAT. The former explores the use of wastewater in agriculture, while BIOFAT aims to develop uses for fats in oils such as soaps and detergents. A third project is ESAPlastics, which aims to test sensor-based perception technologies to detect plastic in shallow water using drones.

Another initiative in which ISEP is involved is the European Project Semester (EPS), which has led to projects carried out by groups of engineering, business and product-design students. One of these, the Aquaponics project, has seen a vertical garden installed over three floors of one building on site, where vegetables are produced for the students and local community. It's one of the best illustrations of the success of ISEP's commitment to sustainability.

"As an institution we are very focused on working with local organisations and working around the campus in order to support activities related to sustainability for ourselves and our neighbouring communities," explains Professor Viamonte. "We have big projects related to waste, plastic and water. These things are very important as we need to be more sustainable on our campus, in our city, in our country and in our world."

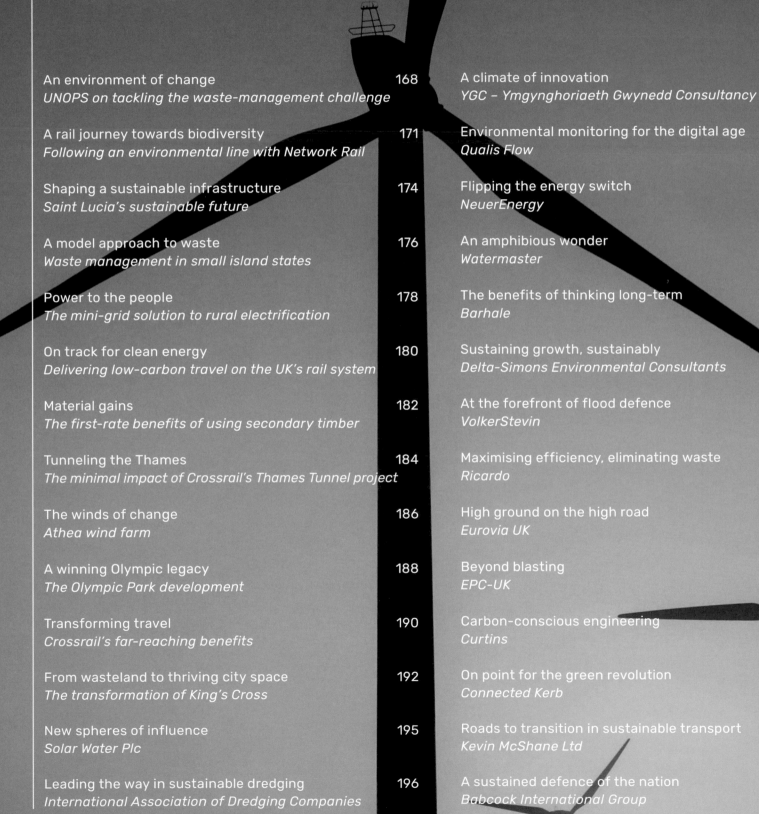

ENVIRONMENT AND ENERGY

An environment of change

UNOPS's recent work in Sri Lanka illustrates how environmental challenges such as waste management call for sustainable infrastructure solutions

The township of Kattankudy in the Eastern Province of Sri Lanka faced overwhelming amounts of garbage and no designated area to dispose of it. With no sustainable plan in place, local authorities disposed of trash in the nearby lagoon and its surrounding area, endangering the residents' health and the environment.

The Urban Council – the local authority in charge of waste management – took action after residents protested. However, initial efforts were difficult to sustain due to expenses or a lack of technical knowledge and operational capacity. With funding from the European Union, UNOPS partnered with the Urban Council to find a long-term solution.

As part of a larger sustainable waste-management programme, UNOPS procured garbage trucks and developed integrated solid-waste management systems in two districts in the Eastern Province that facilitated improved waste-management services for more than 650,000 people.

Income-generation activities also improved the economic sustainability of the programme. The programme also included community education, training and capacity development. In Kattankudy, around 45,000 people were informed about the importance of proper waste handling and the concept of "reduce, reuse and recycle" through public discussions, house visits, campaigns and training events. Educational programmes were also implemented throughout schools. Across the township, 80,000 residents are now better informed on their environmental responsibilities.

Sustainable growth requires minimising the resources used and pollutants generated throughout the extraction, production and consumption processes. These rely on a continued supply of resource inputs, including energy and water, as well as the treatment of waste outputs to the air, water and soil. Redesigning these processes to minimise environmental impact will involve fundamental changes to the way infrastructure systems are planned, constructed and used, and their materials recycled.

Preserving marine and terrestrial ecosystems from environmental degradation, climate change and pollution requires targeted efforts to protect, restore and promote conservation and sustainable use. As the Sri Lanka waste-management programme shows, this requires action across infrastructure sectors.

INFRASTRUCTURE AND THE ENVIRONMENT

Suitable management of solid waste and wastewater can eliminate contamination or pollution, while digital communications technology can provide early-warning capabilities and support countries in understanding disaster risk and strengthening responses through data collection, research and awareness-raising.

Digital information systems also provide useful resources to increase efficiency, raise general sustainability awareness and promote behavioural changes such as reduced energy consumption and recycling. For example, as a part of the Sri Lanka sustainable waste-management programme, UNOPS teamed up with a mobile service provider to establish an internet-based waste collection tracking system to monitor the waste collection fleet.

Kattankudy in Sri Lanka faced serious waste-management issues that called for a long-term, sustainable solution

"Preserving ecosystems requires targeted efforts to protect, restore and promote conservation and sustainable use"

All infrastructure sectors must be designed to minimise their impact on the environment through consideration of how the sector interacts with other sectors and the materials used. For example, the transport and energy sectors are critically interdependent; improved fuel efficiency in vehicles and the electrification of transport can reduce energy demand and fossil fuel emissions.

Given the vital role of energy as a resource for development, infrastructure needs to address core dimensions of energy sustainability, defined by the World Energy Council as the "Trilemma" of energy security, universal access to affordable energy services, and environmentally sensitive production and use of energy.

The design of energy, water-supply and waste-management systems should, in the event of a disaster, allow affected communities to continue to receive basic services, including fuel and adequate sanitation, which will reduce the incidence of epidemics and other social risks.

According to a 2014 IPCC report, infrastructure development is responsible for 76 per cent of man-made global carbon emissions. This carbon is not only from energy use but is also embodied in the extraction of materials for use in construction, the construction methods used and final energy-use requirements for operation.

These demands have a substantial impact on the amount of carbon emitted into the atmosphere, and decisions should be made to reduce them. For example, the embodied carbon in those materials should be reduced to minimise emissions by using less carbon-intensive materials (such as wood instead of steel) or by reusing materials from decommissioned assets (such as rubble as aggregate for concrete walls and pillars).

Finally, conservation efforts require participatory governance through international cooperation. Infrastructure planning across all sectors should

be integrated in national policies to ensure the most vulnerable communities do not suffer disproportionate impacts from climate change. Digital communications can support the knowledge transfer and information sharing necessary to link implementation and policies between international partners.

Climate change presents the single biggest threat to development. To combat its effects and build resilience to natural disasters, SDGs 12–15 call for action to protect the environment and to encourage more sustainable patterns of consumption and production.

A rail journey
towards biodiversity

Network Rail is ensuring that the management of its vast track system goes hand in hand with a considered environmental and community approach

SDGs: 11, 13, 14, 15

The British railway network is the oldest in the world and currently covers more than 50,000 hectares, extending to 32,000km of railway track. Not only does it feature marvels of engineering like the Forth Bridge in Scotland or the Box Tunnel underneath Wiltshire in England, but it also carries over 25,000 trains every day, travelling at speeds of up to 200 kilometres per hour. The railway network passes within 500 metres of one-third of the population of Britain, and 7 million are even closer neighbours, with the railway tracks only 12 metres or less from their back fence.

The management and maintenance of this almost 200-year-old rail network is key to the safe and timely passage of almost 2 billion passenger journeys every year, and it is this work that has often come into conflict with the railway's other neighbours living on and in its infrastructure – the plants and animals that live on the Network Rail estate. The railway's network passes through every National Park in Britain and hosts more than 200 Sites of Special

Scientific Interest (SSSI) – legally protected areas designated for particular habitats and species.

The network also features many important geological formations discovered by those Victorian engineers who blasted their way through the countryside in the 19th century. These SSSIs create a particular challenge as their protection can restrict the maintenance activities that are allowed; work must not have a negative impact on habitats and Network Rail is also legally bound to work to improve their condition.

UNIQUE HABITATS

Ever since the railways were built, there has been legal obligation to provide suitable barriers to access, originally to stop workers trespassing on adjacent land but now to protect people and livestock from fast-moving trains and electrical power systems. Over the past two centuries this restriction to access, together with the varied approaches to vegetation management, has resulted in unique

track-side habitats, and some particularly rare plants can be found, such as the Deptford Pink flowering plant, in a few locations in Devon and Leicestershire.

The railway has avoided using fertiliser, as is common on agricultural land, benefiting the soils and associated species that are found there. The management of vegetation on the estate has evolved over time, from regular cutting of grassland during the age of steam to reactive management of woodland alongside the diesel and electric trains.

The presence of trees next to the railway has the potential to negatively impact both safety and performance, creating hazards and delays for trackworkers, train drivers, passengers and level-crossing users. Trees growing in the wrong place can block signals and sight-lines for level-crossing users. They can also fall onto the track during extreme weather events, blocking the line and causing delays. Hence, recent vegetation-management practices have focused on managing Network Rail's trackside woodland for safety and performance purposes, often to the detriment of valuable habitats such as species-rich grasslands.

VEGETATION MANAGEMENT

Recognising the increased risk to safety and performance from a reactive vegetation management regime, Network Rail sought to create vegetation alongside the tracks that could enable trains to pass safely. This involves trying to manage trees to ensure that they are shorter than the distance from the tree to the track. Often, however, because of the thin linear layout of trees along the rail track, any removal of the 30- or 40-year-old trees on the fence line exposed the railway to adjacent communities. Sometimes Network Rail failed to demonstrate that it had taken either the residents or biodiversity into consideration, resulting in widespread criticism and government ministerial intervention.

An independent review (2018's Varley Review) recommended changes to Network Rail standards and procedures, and also set a target for no net biodiversity loss by 2024 and a biodiversity net gain by 2040. In accepting these recommendations, Network Rail wants to achieve that net gain by 2035. Essential safety vegetation assessments will take place as before, but will also take into account how to protect and nurture the habitats and species that live on the railway, and better consider the impact that vegetation works have on track-side neighbours and local communities.

In response to the Varley Review and the Department for Transport's new policy, Network Rail has launched the Sustainable Land Use programme to transform the way that it values and uses land. The aim is to create a railway landscape that includes running a safe and reliable railway, protecting the environment, enhancing biodiversity and being seen as a responsible landowner and neighbour.

Shaping a sustainable infrastructure

Saint Lucia's policymakers are applying a holistic,
cross-departmental, long-term approach to the
island's extensive infrastructure plans

SDGs: all

Nestled in the middle of the Windward Islands chain of the Caribbean, Saint Lucia is a paradise for tourists. The island draws more than 1.2 million visitors to its shores annually, by plane and ship, to enjoy its sandy beaches, lush rainforest and stunning scenery. However, Saint Lucia faces major challenges providing the infrastructure necessary to support these visitors in the future, in addition to the needs of its growing population and rapidly developing economy.

Water shortages are common, especially in the dry season. Electricity is costly and generated by fossil-fuel burning. Most wastewater is left untreated and landfill space is running out. Roads are jammed with traffic. In addition, the island's infrastructure, and the houses, schools, hospitals and other facilities that it supports, is increasingly at risk of damage or destruction from extreme weather events.

"Saint Lucia is highly vulnerable to the extraordinary changes that are being recorded to our global climate," says Prime Minister Allen Chastanet. Indeed, the increasing intensity of hurricanes, flash flooding and other climate-related risks poses a serious threat to existing infrastructure on the island and the livelihoods of its residents. "We must act swiftly to ensure preparedness for the worst-case scenarios, and to ensure resilience for our people and communities," says Chastanet.

AN ISLAND-WIDE ASSESSMENT

A new National Infrastructure Assessment report provides a key first step in the island's journey toward sustainable and resilient infrastructure, aligning the country's infrastructure planning with its national development priorities and its international commitments under the Sustainable Development Goals and the Paris Agreement. It offers a detailed and comprehensive island-wide analysis of the island's infrastructure systems to facilitate future planning for the country's policymakers. Produced by UNOPS and the University of Oxford-led Infrastructure Transitions Research Consortium, in partnership with the government of Saint Lucia, the report uses an evidence-based approach to anticipate the island's future infrastructure requirements.

"The interdependent modelling tool used to conduct the analysis brings together data from a number of sectors," says Oxford researcher Daniel Adshead, lead author of the assessment report. "It allows users to decide how much of each type of infrastructure may be required on the island in the future, and the kinds of investments, policies and projects that can be implemented in order to meet these needs." In national governments, where decisions on infrastructure sectors are often taken in separate "silos", this approach allows specialists to plan integrated solutions that can be more efficient and cost-effective in the long-term. Given the cross-sector impacts between energy, transport, water and waste management, policymakers should prioritise solutions that can benefit the infrastructure system as a whole.

In addition, the National Infrastructure Assessment emphasises the geospatial considerations of infrastructure planning in a small island context. A flooded substation may interrupt power to numerous homes as well as critical facilities such as hospitals. A storm-damaged road may limit access to schools, airports or markets, putting social and economic development at risk. The analysis uses novel mapping methods to intersect destructive hazards with building footprints and networks so that the potential impacts of a storm, flood, landslide, or sea-level rise can be estimated. As a result, decision-makers can more fully account for resilience aspects when planning new buildings or facilities, and how they might contribute to achieving the SDGs.

"Saint Lucia is to be commended in adopting this approach, which will help ensure that limited resources are applied to maximise socio-economic development, while protecting the country's unique natural resources," says Grete Faremo, UNOPS Executive Director. With expansion of its tourism sector underway – a $175 million airport redevelopment, proposed new cruise terminal, and new hotels and resorts planned around the island – state-of-the-art infrastructure planning tools will be critical to ensuring the island's growth is conducted with sustainability and resilience in mind. Furthermore, it can act as a regional pioneer and an example to its neighbours. As Faremo explains: "This is an approach that is equally relevant and increasingly vital for other Small Island Developing States during the next decade."

OPPOSITE

The recent National
Infrastructure Assessment
report puts an integrated
approach at the heart of
Saint Lucia's plans for
sustainable growth

A model approach to waste

The Environmental Change Institute at the University of Oxford is working with Curaçao's government and industry to develop alternative waste-management strategies

SDGs: 9, 12

Confronted by growing numbers of tourists and expanding residential populations, small islands such as Curaçao in the Caribbean are faced increasingly with unmanageable volumes of solid waste. Zita Jesus-Leito, Minister of Traffic, Transportation and Urban Planning, Government of Curaçao, echoes this sentiment. "Tourism is an important pillar of our economy," she says. "But more tourists also means more waste and more pressure on our already overstretched infrastructure."

Lena Fuldauer from the Environmental Change Institute at the University of Oxford agrees. "In addition to landfill capacity constraints and a large problem with illegal dumping, new types of wastes are being brought to small islands, which cannot yet be treated sustainably," she says. "The unfavourable economics of recycling in such small countries exacerbate the waste management problem."

Many small islands are faced with waste problems. It is widely reported that limited waste-disposal space, underdeveloped infrastructure, lack of regulation, and poor enforcement have stalled progress on the sustainability of the sector. The reasons for these problems are complex, but include an absence of target setting and visioning for sustainable outcomes, a lack of appropriate data collection, and limited technical and institutional capabilities.

DELIVERING SUSTAINABLE OUTCOMES

However, these problems also provide an opportunity to use waste-management planning to leap-frog directly to deliver sustainable outcomes – in alignment with the Sustainable Development Goals. Studies have argued that integrating a range of stakeholders throughout the process of waste-management planning can assist in data collection and identification of options with local buy-in.

Together with the government of Curaçao and the United Nations Office for Project Services (UNOPS), a team of researchers from the University of Oxford has developed novel modelling capability to improve sustainable waste management. Fuldauer explains: "Data analysis and modelling of waste management scenarios helps identify different technology and

policy options; implementing these options supports sustainable development into the future".

However, this is not easy. Appraising the performance of such options under uncertain future scenarios requires a formalised and systematic modelling tool. This is where the open-source modelling software developed at the University of Oxford comes in. The modelling capability was successfully used in the small island of Curaçao, where the project team engaged with more than 20 national decision-makers from government and industry, to develop alternative waste-management strategies.

The project team simulated the performance of various waste-management strategies under a range of plausible futures with this open-source modelling software for Curaçao. Using inputs from local stakeholders, the model shows that the immediate introduction of sustainable waste-management initiatives can deliver on all waste-related Sustainable Development Goal indicators, while delaying landfill depletion. Prevention campaigns provide the lowest-cost means of dealing with the waste problem, while inter-regional recycling projects have the potential to tackle the waste problem by directing large quantities of material away from the landfill. Investment in such low-regret sustainable waste management initiatives now can reduce the potential for future greenhouse gas emissions and an economic dependency on large-scale technologies.

The open-source data and model used in this study was made available to government, researchers and the private sector in Curaçao. The team also held an extensive training workshop to ensure that people could use it. This enabled decision-makers to evaluate the consequences of actions before committing to them, thereby helping to depoliticise investment spending.

Minister Jesus-Leito endorsed the research. "With the help of UNOPS and the University of Oxford, we deepen our understanding of how evidence can be used in decision-making to best transition infrastructure as a driver of sustainable, social and economic development," she says. "The evidence-based approach has provided practical solutions on how we can address our challenges, for example waste."

ABOVE
Expanding tourist and
resident numbers are putting
the waste-management
systems of many small islands
under extreme pressure

ENVIRONMENT AND ENERGY   **177**

Power to the people

Mini-grids – with their own power-generation system and local distribution network – can address the challenge of delivering electricity to rural parts of sub-Saharan Africa

SDG: 7

In 2019, around 840 million people worldwide had no access to electricity, with around half of those living in rural areas of sub-Saharan Africa (SSA). By 2030, the World Bank estimates that 650 million will be without access to electricity, with 90 per cent of them living in SSA. Rural communities without electricity become locked into a subsistence lifestyle, which exacerbates the rural–urban poverty gap and reduces their quality of life. Extending the electricity grid to dispersed and hardest-to-reach communities in SSA has historically been very slow, with repeated failures to meet ambitious national connection targets.

Mini-grids, encompassing a power generation system and a local distribution network, have the potential to supply electricity to remote areas of SSA where the required investment to extend the national grid is prohibitive. To date, such mini-grids have often been powered by diesel generators, creating a supply chain maintenance weakness and pollution. Diesel generators represent a small capital investment with an apparent advantage over renewable energy systems such as photovoltaics (PV), where the fuel is free, but there are high upfront capital costs. Agencies can deploy a far greater number of diesel-based mini-grids for a given budget. However, there is now a growing recognition that, in addition to environmental concerns, such an approach undervalues the maintenance and reliability advantages of PV over diesel, especially in remote, hard-to-reach areas.

In addition, since 2010, utility scale solar PV has seen a 75 per cent cost reduction, while utility scale batteries have halved in price over the same period. As a result, the capital investment barrier, while still there, has narrowed dramatically. When combined with the inherent modular nature of PV and batteries, such technologies are now seen as very appropriate in SSA mini-grids.

MINI-GRID PROGRAMME

Technically, mini-grids do not generally represent a particularly difficult design challenge in an SSA context. The solar resource is excellent, with all-year-round availability – their success (or failure) is down to the level of understanding of the deployment context, the cost of electricity and their operational management. Over the past 10 years, the University of Southampton's Energy for Development (e4D, www.energy.soton.ac.uk) programme has been looking at community-based mini-grids in Kenya and Uganda. "The basic premise of e4D is that productive use of electricity alongside low-power appliances have the ability to transform the economics of off-grid electricity provision," says Professor Bahaj, who leads the programme. "Furthermore, technology innovation such as mobile money means that there is both a need and a willingness to pay for electricity services among even the poorest of society."

The e4D approach connects buildings within a village trading centre. Shipping containers are converted on site to form the basis of the e4D plant room, energy centre and structural support for the PV canopy. A village energy cooperative is responsible for the day-to-day operation of the system (maintenance, electricity sales), creating local employment alongside strengthening village "identity and ownership" of the mini-grid. Across the mini-grids deployed in five villages to date, consumers tend to overstate their level of turnover prior to the mini-grid deployment (to appear successful) which leads to the risk of system oversizing.

Affordability of electricity, in comparison to the subsidised utility grid, is perhaps the biggest barrier to mini-grids in SSA. Villagers expect the utility grid price even when it is clear that the subsidised utility grid will not arrive. Such mini-grids therefore should ideally be deployed in partnership with a country's rural electrification authority to ensure a level financial playing field so that the poorest in society can benefit from the transformative effects of electrification.

As the e4D programme goes to show, mini-grids can provide more reliable and sustainable electricity, even far off-grid, with benefits for local communities, supporting new jobs and the rural economy.

ABOVE
Network Rail put low-carbon
rail travel at the heart of its
plans to rebuild Birmingham
New Street station

On track for clean energy

Network Rail and a host of partners have been working to make low-carbon rail travel in the UK a reality

SDG: 7

The UK rail system is one of the country's most low-carbon forms of transport, but Network Rail still has a vital leadership role to play in minimising its impact on the environment, from developing a new way to power trains to investing more than £650 million in clean energy for a major station.

The question of whether trains can be powered directly from solar panels, for instance, is an apparently simple one with great potential, given that there are many locations that are suitable for solar farms but where a grid connection would be prohibitively costly.

However, significant research and piloting was needed to demonstrate that a solar farm could safely manage the intermittent flow of electricity when a train accelerates out of a station. A single train can require up to 4MW – enough energy to power 2,000 homes for an hour – to do so, meaning that the power connection needs to be very robust to prevent the risk of dangerous electrical faults or fires.

After extensive collaboration between the community energy charity Riding Sunbeams and Network Rail, First Light – the world's first solar farm connected directly to traction power substations – started generating electricity in Aldershot, Hampshire in June 2019. As a demonstrator, its purpose is to gather evidence to enable the rail industry to redesign the DC power supply system. As Network Rail currently uses around 1 per cent of the electricity generated in the UK, if adopted, it would be a game-changer for decarbonising trains.

"This would never have happened without some key champions within Network Rail having the vision to see that we were onto something really exciting, and the courage and tenacity to help us push through the technical and administrative barriers to make it a reality," says Leo Murray, Director of Riding Sunbeams. "Thanks to this work, direct supply from renewable generators will now play a key role in plans to decarbonise all trains on the network, with major commercial potential for rollout of this innovation worldwide, too."

BIRMINGHAM NEW STREET

In another example of Network Rail's innovative approach, between 2010 and 2015, it invested £650 million in the rebuilding of Birmingham New Street station to make low-carbon rail travel more attractive. By collaborating widely and using a recognised green methodology (BREEAM) for the first time on the UK railway, the project team was able to make dramatic savings in energy, water usage and waste.

The first ever Combined Heat and Power (CHP) plant to power a UK station was installed. CHP is highly efficient, as it uses the often-wasted heat from power generation to heat buildings. However, the station requires eight times more electricity than heat, so the project teamed up with the Birmingham District Energy Scheme to share heat across a range of Birmingham's landmark buildings via 1.5 km of insulated pipework, saving more than 15,000 tonnes of CO_2 emissions each year. Solar power and highly efficient air-source heat pumps also power railway staff offices on Platform 1.

Birmingham New Street's water usage was reduced by 60 per cent by using rainwater from the roof to flush the station toilets, and 98 per cent of all construction waste was recycled or reused to divert waste from landfill. The station is now flooded with natural light from the lightweight roof, including light wells that serve the covered platforms below the main concourse. In addition, low-energy LED lighting across the station uses infrared sensors to dim or turn off the lights when they are not in use.

The project has been a great success, with Birmingham New Street now serving 47 million passengers each year and ranked consistently in the top five stations for passenger satisfaction.

Material gains

A new process for reusing timber could help reduce the construction industry's heavy environmental impact

SDGs: 8, 11, 9, 12

The construction industry is responsible for more resource use and waste generation than any other industry. Greenhouse gas emissions from the production of new steel and concrete alone amount to 10 per cent of the global total. The search is on to find better ways to make use of materials available in end-of-life buildings, which currently leave the industry as waste. This reduces the impacts of burying, burning or downcycling old materials, and can also reduce the mining, quarrying or logging, and processing and transport of new materials.

Conventional cross-laminated timber (CLT), made of virgin softwood, has become established as an alternative to steel and concrete over the past two decades. It consists of timber planks, stacked and glued in several perpendicular layers into panels that can be as long as a transport lorry. These can be pre-cut using computer-controlled cutting equipment for efficient assembly on site to provide the primary structural frame of buildings. CLT buildings go up fast and are approaching cost parity with other construction methods.

Andrew Waugh of Waugh Thistleton Architects is one of the pioneers of this movement. "CLT building sites are happy, quiet and clean," he says. "You don't have the noise of jackhammers, grinders and cement mixers because essentially you are putting up prefabricated parts with cordless screwdrivers."

Timber is one of the few renewable materials widely used in construction. Putting timber into the built environment is a way of storing the atmospheric carbon dioxide that a tree sequesters. And because buildings last a long time, that carbon is locked up over the long term.

SOMETHING OLD, SOMETHING NEW

Nevertheless, timber is not an infinite resource, and comes from distant forests with an important role for biodiversity. Since timber is also in demand as a renewable alternative for other resources,

such as fossil fuels and plastics, making the most of the timber that we harvest is essential. Cross-laminated secondary timber (CLST), which extends the life of timber that is reclaimed from demolition, is an innovation in its early stages. CLST exploits the residual strength and beauty of solid wood, instead of resorting to chipping it – for recycled products such as MDF or chipboard – or incinerating it for energy.

"Finding ways to reuse timber that is already a part of the built environment makes a lot of sense," says Dr Colin Rose, lead researcher and inventor of CLST. "Timber from old buildings is often of a better quality than new wood, and the longer we keep it in use, the more it contributes to locking up carbon." However, putting reclaimed materials straight into new buildings presents difficulties for an industry that is geared around consistent, certified new products. Reclaimed materials tend to arise in relatively small, mixed batches, without warranties; often they do not command a high enough price to justify the cost overheads of sorting, storing and reselling. "Transforming secondary timber into CLST presents a business case for reusing or upcycling materials on an industrial scale," says Dr Rose. "The process turns low-value materials into a standardised component. It's the kind of product we urgently need to meet the demands of the construction industry, while vastly reducing environmental impacts."

For CLST usage to reach its full potential, producers require sources of secondary timber that are close to where new materials are needed. Cities present this opportunity. The solid timber waste from a city could be diverted to dedicated reprocessing plants, creating more employment than current means of recycling and disposal. "This is an innovation with the potential to create benefits on many fronts – environmental, social, and in contributing to circular local economies," says Professor Julia Stegemann of University College London, who leads the ongoing research to support the implementation of CLST.

Tunnelling
the Thames

London's Crossrail project crosses the Thames at only one point, and the construction team made sure it did so with minimal environmental and social impact

SDG: 12

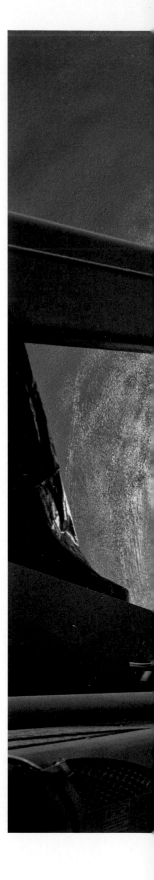

Protection of the environment is a core component of Crossrail's sustainability strategy, including protection of natural resources, minimising the impact of noise, vibration and emissions from site operations, and reducing the project's carbon footprint.

The Thames Tunnel project involved the construction of two tunnels (and portal structures) under the River Thames between Plumstead and North Woolwich, totalling 6km in length. It is the only part of the Crossrail route that crosses the Thames. The £260 million tunnel project was highly complex with a distinct set of challenges, such as a large head of water above the tunnel's deepest points and having to tunnel through chalk ground. In addition, being adjacent to busy roads, listed buildings, a live railway and residential properties presented space-constraint challenges that needed to be overcome.

Crossrail promoted sustainable design by mandating the use of CEEQUAL for the design and construction of the tunnels, portals and shafts, and BREEAM Bespoke sustainability assessment for underground stations. It created a degree of "competition" between teams to achieve the best possible sustainability score. This helped to drive the best possible outcome and encourage innovations within the overall project constraints. It also helped maintain commitment and motivation across the supply chain.

SUSTAINABLE METHODS

The tunnelling, carried out with mixed-shield, or "slurry", tunnel-boring machines (TBMs), was located in the chalk layer and required extensive dewatering. It was not possible to put all the water removed back into the ground so, rather than discharge it into the sewer system, the site team implemented a scheme to use the surplus groundwater for bentonite mixing, slurry dilution and grout mixing, saving over 2 million litres of potable water a week. The use of groundwater for these processes is a new and innovative practice.

An overall reduction of 11 per cent in carbon emissions was achieved across the life of the project. This was done with key innovations such as saving water, using ultra-efficient and hybrid plant machinery and introducing diesel particulate filters, which were retrofitted on all machines.

The project required significant quantities of concrete (more than 250,000 tonnes), steel (more than 7,700 tonnes) and granular fill (more than 82,500 tonnes). By using high-performing replacement concrete, 100 per cent recycled steel and recycled granular fill, it achieved an overall recycled content in the materials used of 35 per cent.

Over the four-year project, the construction team actively sought solutions to reduce the environmental impact of the materials used. Innovative design changes were suggested by the contractor pre-construction and implemented by the designer and client, resulting in significant savings of resources, waste, embodied carbon, cost and time.

The contractor employed an unusual technique for launching the two TBMs. Instead of employing the traditional technique, around the eye of the tunnel, the team installed a re-usable steel frame that could be dismantled and moved to facilitate the launch of the next TBM. This removed the need for the large concrete blocks (200–240 tonnes) and steel frames (300–320 tonnes) usually needed that could not be re-used. The project also reduced the noise associated with breaking them out, and the associated lorry movements. This also removed the need for a total of 20 sacrificial tunnel rings (500 tonnes of concrete and steel fibres) during the four launches.

Large, heavy tunnel rings were used to construct the tunnel. The contractor proposed an innovative change in tunnel ring design that reduced the number of tunnel rings by 7 per cent. Due to the size and weight of the tunnel rings only one could be transported to site per lorry. With 7 per cent fewer rings required, this took more than 250 lorry journeys off the roads.

This project demonstrated technical innovations driven by a collaborative team approach to deliver maximum sustainability performance. The project's sustainability credentials were confirmed by achieving a CEEQUAL Excellent (89.1 per cent) Whole Project Award rating. The project also won the Water Resources category award at the CEEQUAL Outstanding Achievement Awards 2016.

Crossrail mandated the use of CEEQUAL
for the design and construction of the
Thames Tunnel project's tunnels, portals
and shafts

The workforce underwent ongoing environmental training thoughtout the wind farm's construction

The winds
of change

Constructing the Athea wind farm in Limerick required
imaginative re-use of earthworks and sensitive treatment
of the local flora and fauna

SDG: 7

The Athea wind farm site in Limerick, south-west Ireland spans
a variety of habitats, including upland blanket bog and heath,
coniferous tree plantations, and rough agricultural grassland.
Building the wind farm involved the construction of access
roads, hardstands and 16 turbine bases. Four existing turbine
bases and a substation building needed to be demolished, and
a permanent met mast was installed (to collect meteorological
data) along with 16 GE-manufactured wind turbines with a total
capacity of 34.35 megawatts. Work on the site began in
September 2012 and was completed in February 2014.

Roadbridge, the civil engineering firm behind the project,
used the CEEQUAL assessment scheme to provide a systematic
and coherent approach to tracking the project's management
activities and following the best sustainability practices.

The wind farm's site lies within a Special Protection Area (SPA),
largely due to the presence of the hen harrier, as well other protected
bird species such as the short-eared owl, the merlin and the red
grouse. Indeed, one condition of the SPA planning permission
was that all earthworks ceased completely between 1 April and
1 September to coincide with the breeding season of the hen harrier.

ENVIRONMENTAL AWARENESS
The project placed a strong emphasis on environmental awareness
for all working on site. During the site induction, all new workers
and sub-contractors were given an eco-map of the site showing
all the eco-sensitive locations, designated refuelling locations
and conservations areas. Throughout the construction phase,
the workforce underwent an ongoing programme of environmental
training. Each morning, all site workers attended a pre-task briefing
where the scheduled tasks for the day were discussed and any
environmental controls outlined.

During the project, several local community outreach initiatives
were put in place, including the donation of labour, equipment
and materials to support the Athea Tidy Towns Committee in the
construction of a new walkway in the local park; a presentation by
the site environmental team to Athea National primary school
pupils on the area's ecology and biodiversity; and a following school
visit to the site when the site management presented the workings
of a modern wind farm.

The team managed to reuse all earthworks material on site, which
is unusual for a project this size. No earthworks material was sent
off-site. A site-specific waste-management plan was prepared and
implemented on this project. The plan set stringent site targets for
waste stream segregation, recycling rates and diversions rates from
landfill disposal, which were all achieved.

At the start of the project, 95 per cent of the structural
components required were brought from a previous Roadbridge
project. The construction of the site compound is modular in
nature, which ensures that it can be completely remobilised to
the next site with no wastage. This is not the norm, but saves
a huge amount of wasted materials and time.

From a greenhouse gas inventory analysis of the project,
there will be a carbon payback period of 2.69 years for this project,
thanks to the mitigation measures employed at planning, design
and construction phases. A green travel plan was produced
following travel-to-work surveys completed by all site users.
A private bus was hired to collect workers from surrounding
villages to travel to work. This saved carbon emissions from
unnecessary private car journeys.

A rainwater-harvesting unit was constructed on-site to capture
rain run-off from the roof of the main compound. This harvested
rainwater was used to supply water for the on-site toilets.
This resulted in an average saving of 34 cubic metres of water
per month. The project, which won the CEEQUAL Outstanding
Achievement Award 2016 for Energy and Carbon, minimised its
environmental impact and left a lasting impression on the local
schoolchildren, plus a green-energy-generating asset to help
decarbonise power generation.

A winning
Olympic legacy

To ensure that London 2012 would be "the greenest games ever", the Olympic Park had to be sustainable, accessible, useful and ecologically sound

SDGs: 3, 9, 11, 12, 14, 15

"From the delivery of the infrastructure to support one of the most memorable events in British sporting history – the London 2012 Olympic and Paralympic Games – to a major regeneration project transforming the setting of the games into stunning parklands and waterways, the Olympic Park development has provided the public with a place to play, relax and explore." So said Dan Whiteley, Head of Environment at BAM Nuttall, the contractors responsible for large elements of the Olympic Park development in east London.

The 2012 Olympic Games aimed to be "the greenest games ever". Over the course of the eight-year preparation programme, London 2012 had five sustainability themes – climate change, waste, biodiversity, inclusion and healthy living. This resulted in the development becoming an exemplar for best practice in environmental sustainability, and ultimately ensured the regeneration of a previously part-derelict, post-industrial brownfield site to a stunning landscaped public realm space.

Back in 2006 the heavily contaminated soils on this land needed a lot of remediation, which was done using a highly innovative on-site soil treatment centre that enabled the reuse of up to 90 per cent of the treated material. Additionally, groundwater and numerous existing contaminated waterways were remediated and the planting of softer embankments enabled the creation of new wildlife habitats.

BAM Nuttall conducted the post-Olympic Games legacy transformation works, a major regeneration project to transform the site into urban parklands and waterways. The task was to clear the games-specific infrastructure (including temporary venues, walkways and roads); to connect the park with new roads, and cycle and pedestrian paths that stretch across the site and into the surrounding area; and to complete the permanent venues, bridges and parklands for their intended future use.

SITE REGENERATION

The key to the regeneration of the site was in making its green spaces more accessible. Vast open areas and flora beds were created across the park with more than 1.3 million plants and one thousand

semi-mature trees. Specialist top soils were designed to encourage the development of new ecosystems, as part of the commitment to a more sustainable future. This diverse range of trees and plant life formed inner city woodlands, wetland areas and extensive wildflower meadows, with all trees and plants in the park carefully selected to ensure that green spaces are future-proofed against climate change.

Much of the transformation work was undertaken in a live recreational environment around planned sporting and entertainment events. Park-wide collaboration in sustainable design approaches across the contractors, sub-contractors and client brought added benefits – efficiency savings (in terms of time and money) and the sharing of best practice. High-quality temporary bridges were designed and constructed to allow for 100 per cent recyclability or re-use upon dismantling, achieving significant waste minimisation, reductions in the use of construction materials and a reduced carbon footprint.

The sustainable design also considered planned housing developments, with the success of the park instrumental in drawing people into the area and creating a desirable location to live. Children's playgrounds were developed around the concept of ecological processes and lifecycle, and used as an educational aid. This involved working closely with local businesses and educational institutions to deliver a range of exciting and stimulating play experiences.

The main Olympic Park concourse plaza, once entirely paved, has been transformed into a vibrant new space, playing host to a wide range of cultural events. It features a wide, tree-lined promenade as a new social centrepiece of the park, which connects a series of exciting spaces designed to inspire creativity in people of all ages.

The Olympic and Paralympic facilities were the most sustainable ever built – on time, on budget and built to high sustainability standards. The resulting Olympic Park is the largest new urban parkland in Europe for 150 years, including attractive new and existing waterways with links to surrounding areas via new highway, cycleway and rail networks.

The Olympic Park development has transformed a post-industrial brownfield site into Europe's largest new urban parkland for 150 years

ABOVE
Crossrail set out to deliver
a world-class railway fit
for generations to come

Transforming travel

London's biggest transport infrastructure project in years, Crossrail aims to deliver significant benefits for generations to come

SDGs: 8, 9, 11, 12, 15

By 2030, the capital's population is set to reach 10 million and its transport system must be able to meet the resulting rise in demand. Crossrail is delivering a new railway to add 10 per cent to central London's rail capacity and to help maintain London's place as a global city.

When complete, the Elizabeth line will offer a high-frequency, high-capacity service linking 41 stations over 100km from Reading and Heathrow in the west to Shenfield and Abbey Wood in the east. It will transform travel in London and the South East.

The net economic benefit of Crossrail is substantial. From the jobs created during construction and operation and time saved by improved transport connections, to the enhanced accessibility the line will bring to millions of people, the project is estimated to generate at least £42 billion for the UK economy.

Crossrail has been a project of huge ambition. In one of the busiest cities in the world, a large, diverse workforce has been delivering 42km of new tunnels, ten new stations, and upgrades at many more, while keeping the city moving. In addition to the rail line, Crossrail is delivering new public spaces around stations, as well as new homes, offices and shops along the route to integrate the new infrastructure into the communities it serves. In addition, world-class public art installations are being created to enhance the experience during travel and reinforce London's place as a thriving centre for art and creativity.

During construction, the Crossrail project has supported 55,000 jobs around the UK. It has trained a new generation of engineers in tunnelling, underground construction and railway engineering through its purpose-built academy. It has also recruited more than 1,000 apprentices, injecting new talent into the industry. The scale of the job has driven innovation, helping to improve safety and productivity across the construction industry. Off-site manufacture at a scale not seen before in the UK construction industry and a custom-built robotic drilling rig have demonstrated benefits in productivity, accuracy and, importantly, safety.

SUSTAINABLE SOLUTIONS

Eight million tonnes of excavated material were generated during construction of the tunnels, shafts, station boxes and caverns. Around 99.7 per cent of this was beneficially re-used to develop new nature reserves, recreational facilities, and agricultural and industrial land. On a tonne per km basis, 80 per cent of this material was transported by rail and water, significantly reducing lorry journeys on the streets of London.

Working in partnership with the Royal Society for the Protection of Birds (RSPB), more than three million tonnes of excavated material was used to create a new 115 hectare intertidal area of saltmarsh, islands and mudflats known as Jubilee Marsh on the RSPB's Wallasea Island Wild Coast project, transforming two-thirds of the island's arable farmland into wildlife-rich habitat. A bird survey conducted on the island identified 29 species and 7,839 individual birds on Jubilee Marsh.

Crossrail has been committed to reducing its carbon footprint by reducing the energy consumption of the operational railway, the embodied energy in construction products and the energy used during construction. A construction-related carbon emissions reduction of 18.6 per cent has been achieved. And reductions in the embodied carbon footprint have been made possible through the use of concrete with cement replacements, with 72 per cent cement replacement achieved at sites such as Paddington station.

Crossrail set targets for weight (350 tonnes unladen weight per 200 metres train) and energy efficiency (24 kilowatt hours per train) for the trains, equating to 55g of CO_2 per passenger kilometre. A train weight of just under 319 tonnes and energy efficiency of around 14 kilowatt hours per train have been achieved, equating to approximately 32g of CO_2 per passenger kilometre, making the new trains among the best in their class for energy efficiency.

The programme set out to deliver a world-class railway fit for future generations and in the process it has made – and continues to make – a positive economic, environmental and social contribution to the city, the country and global industry.

From wasteland to thriving city space

The King's Cross St Pancras project created of one of Europe's largest urban spaces, while taking local environmental and heritage factors into account

SDGs: 8, 11, 15

The regeneration of King's Cross was one of the largest and most challenging redevelopments of landscape and public amenity space in Central London for many years. A 67-acre industrial wasteland was successfully transformed into 8 million square feet of mixed-use space including new homes, shops, offices, bars, restaurants and a university, all while minimising impact on the commuters and pedestrians who passed through King's Cross St Pancras Station every day.

The transformation has provided a new heart for the district and a place for visitors to sit and enjoy, away from the main pedestrian flows of King's Cross. "The King's Cross St Pancras project was based at one of the UK's busiest train stations and London's busiest underground station," says Dan Whiteley, Head of Environment at the project's contractors, BAM Nuttall. "As such, it involved consistent communication with key stakeholders and the local communities to successfully transform wasteland into attractive new city space."

King's Cross has a rich industrial heritage and so the works were carried out in consultation with Historic England. Significant logistical challenges needed to be overcome as construction works bordered the railway and were subject to Network Rail approval. A demanding timeline was needed to meet the official opening of the new St Pancras Eurostar terminal by Her Majesty The Queen.

At the Eastern Goods Yard, land was prepared for the erection of new buildings, with the remediation of contaminated spoil, and hard and soft landscaping. A bridge over the Regents Canal was removed and reconstructed. New roads were built, new utility services installed, redundant buildings demolished, and spectacular fountains were created in the new Granary Square.

The Eastern Goods Yard gas works site south of the Regents Canal was heavily contaminated with hydrocarbons. A bespoke remediation plan was developed, including the installation of dust and vapour monitors, along with odour suppression units to meet the strict requirements for air quality. Extensive ground investigation work was carried out with chemical analysis to inform the remediation strategy, and materials identified for re-use were

ABOVE
The transformation of King's Cross
has created a thriving new space for
locals and visitors alike

managed through the Contaminated Land: Applications in Real Environments (CL:AIRE) code of practice. A lot of hazardous material from site was removed; some of it was treated and reclassified, with bespoke remediation strategies deployed for each of the Grade II-listed gas holder structures.

HERITAGE FEATURES

The site is a historic railway goods yard dating back to the 1850s. The project leaders worked closely with Historic England and the London Borough of Camden to retain historical features including the granite setts and rail tracks. Two 16-feet turntables, constructed in the 1800s, were carefully removed and restored. Extensive archaeological investigations were carried out across the site including documenting the Granary Basin, Granary Building tunnels and stables, Stanley Buildings and the Imperial Gas Light and Coke Works. Now there is a cobbled plaza of 8,000 square metres stretching from the college's Granary Building down to the Regent's Canal.

Early on in the project, part of the Grade II-listed Stanley Buildings had to be demolished. Many materials and fittings,

including fireplaces and doors, were salvaged from the building, and were stored and reused in the refurbishment of one of the other buildings. Ultimately, no material was sent to landfill and many important heritage features were retained for posterity.

The project achieved the Mayor of London's top award for Planning Excellence at the London Planning Awards. "King's Cross is a stunning development that embraces the past while looking to the future," said the then Mayor of London, Boris Johnson. "The transition from a depository of Victorian grain, to a university where the seeds of artistic ideas sprout, has been handled with great sensitivity and flair. It has brought an enlightened focus to the wholesale regeneration of the surrounding area."

Historically, the land to the north of King's Cross was undervalued and underwhelming to visitors. Now it has been transformed into one of the largest urban spaces in Europe, forming the focal point for the district. The project required consistent communication and engagement with key stakeholders and the local community, and ultimately, the six-year project was delivered sustainably and on time.

New spheres
of influence

SOLAR WATER PLC
Location: London, UK
www.solarwaterplc.com

People have been turning seawater into freshwater for years with desalination plants, which often use the process of reverse osmosis. This requires the use of fossil fuel to push seawater through membranes, which can have very adverse environmental consequences for land, sea and air. Malcolm Aw felt there had to be a better way to harness solar energy. After three years of research and development with Cranfield University, a world authority on concentrating solar power technology and water, he created Solar Water Plc.

The solution takes the form of a sphere, the foundation of which is made of concrete and buried in the ground. A steel cauldron is placed inside and a steel-framed glass dome fitted over the top. This is surrounded by concave mirrors that focus the power of the sun and heat up the sea water that is pumped into the dome, during which process the salt is removed and subsequently sold.

"A 50-metre dome will produce approximately 13 million cubic metres of water per annum," says CEO David Reavley. "The water can be used for agricultural purposes – it can be used to irrigate acres of land to grow fruit and vegetables. It's also aesthetically

pleasing, rather like the Eden Project in Cornwall. If you wanted to build a five-star hotel on a beach, the last thing you want to see is a standard desalination plant – this is far better. It satisfies so many needs: agriculture, industry, tourism, as well as drinking water."

Since finalising the design, the Solar Water team has been busy travelling the world to sell their dome and expect to start construction soon. Clients are located in Asia, the Middle East, the Caribbean, the Far East and even some of the southern parts of the United States, and interested organisations include business, water authorities and municipal bodies.

"More than 4,000 children die every day because of water shortages and 844 million people have no daily access to water," says Reavley. "Global water shortages are predicted in the very near future. This is a sustainable solution that requires no fossil fuels. It is cheaper and quicker and easier to build and maintain than other desalination solutions, so the water is less expensive even before you factor in the environmental issues. We can make a real difference to peoples' lives at significantly less financial and environmental cost."

Leading the way in sustainable dredging

INTERNATIONAL ASSOCIATION OF DREDGING COMPANIES
Location: Worldwide
www.iadc-dredging.com

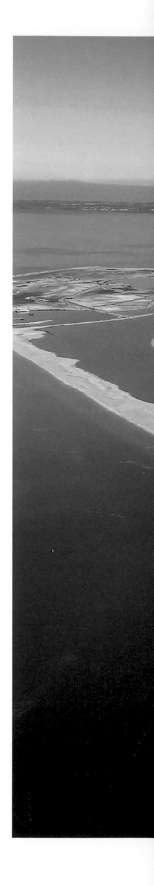

Established in 1965, the International Association of Dredging Companies (IADC) is an organisation comprised of private dredging companies and has 13 main members and more than 100 associated members. IADC is dedicated to promoting the skills, integrity and reliability of its members as well as the dredging industry as a whole. The association spearheads diverse activities and projects to educate, excite and engage its members with an international audience within the dredging sector, its related industries and beyond. By informing the public that the dredging industry is the frontrunner of sustainability.

Dredging is but one component of an infrastructure project, and any one piece of infrastructure functions as a part of a larger network of infrastructure as well as the surrounding ecosystem. The word infrastructure is used to refer to the diverse range of structures, features and capabilities that are developed through the use of dredging such as navigation channels and waterways, ports and harbours, levees and dykes, as well as nature-based infrastructure such as beaches and dunes, islands, wetlands and many other forms of habitat. The association promotes sustainable activities among the dredging industry and the public, and presents radically different methods to address the increasing climate pressures on low-lying deltas as well as modern-day society's incredibly strong demands on the sustainability of water infrastructure projects.

Through its joint publication with the Central Dredging Association, *Dredging for Sustainable Infrastructure*, guidance compiled by a team of scientists and practicing industry experts is made available to project owners, regulators, consultants, designers and contractors. The message is that, through a thorough understanding of economic, environmental and societal systems, and with proactive engagement of stakeholders throughout a project's phases, a value-added water infrastructure project can be successfully achieved. Multidisciplinary project teams must adopt entirely new ways of thinking, acting and interacting. "Knowledge of these solutions has been tailored into a course where experienced lecturers will inform about the latest thinking and approaches, explain methodologies

and techniques as well as demonstrate – through numerous practical examples – how to implement this information in practice with challenging workshops and case studies," says Rene Kolman, Secretary General of IADC.

The association acts as a knowledge disseminator, with its own quarterly print publication about maritime innovations. The quarterly print journal published by IADC, *Terra et Aqua*, continues to feature articles about projects that successfully implement the principles and methods presented in *Dredging for Sustainable Infrastructure* to realise sustainable infrastructure. Articles included projects aligned with the book's philosophy including the Marker Wadden, Harlingen Mud Motor Pilot, and the Prins Hendrik Zandijk in The Netherlands as well as the UK's first sandscaping project in Bacton, Norfolk. The journal's articles as well as all other publications and information about IADC dredging seminars and courses can be found in the "Knowledge Centre" on the IADC website, a digital and searchable library of information about the dredging industry.

As a means for developing water resources infrastructure, dredging relates to each of the SDGs to varying degrees. The use of dredging to construct efficient and productive navigation infrastructure is directly connected to SDGs 2, 7, 8, 9, 10, 11, 12, 13, 15 and 16. As a tool used to provide coastal protection and infrastructure supporting flood risk management, dredging clearly supports SDGs 1, 3, 7, 10, 12 and 14, among others.

In the future, one of the challenges that should be addressed by the dredging and water infrastructure community is to incorporate these goals into the infrastructure development process and to become effective at communicating how such projects support the SDGs. In March 2019, at the Fourth Session of the UN Environment Assembly (UNEA-4) in Nairobi, Kenya, a workshop about *Development of Sustainable Water Infrastructure* was given by dredging industry experts from IADC and CEDA. This workshop enabled cross-pollination to an audience of decision-makers from other industries while advancing the message of sustainable dredging to their respective sectors.

One of the largest nature restoration projects
in Western Europe, the Marker Wadden is a
bird paradise that consists of a 1,000-hectare
landscape above and below the waterline

A climate of innovation

YGC – YMGYNGHORIAETH GWYNEDD CONSULTANCY
Location, Gwynedd, Wales, UK
www.ygc.wales | www.ygc.cymru

Civil engineers are working at the forefront of climate change and nowhere is this more apparent in the UK than in the Welsh coastal village of Fairbourne. There the 677 residents are facing a very real threat from rising sea levels, and this is where Ymgynghoriaeth Gwynedd Consultancy, or YGC, comes in.

"We are working with Natural Resources Wales towards getting the village to a place where it will be possible to facilitate relocation when the defences are no longer sustainable," says Rob Williams, YGC's water and environment service manager.

It's a unique problem for a unique company. YGC was formed in 1996 by the local authority as a way of retaining local engineering talent while delivering civil engineering projects. It now competes with the private sector, tendering across Wales.

Gwynedd has almost 200 miles of coastline to defend from rising sea levels, and works to implement the Shoreline Management Plan 2 that identified communities at risk.

While some of these communities are defendable, it is unsustainable to construct more and higher flood defences in others. Fairbourne is in the latter category and the Fairbourne: Moving Forward project has focused on raising public awareness and establishing a platform of communication with the community for the decommissioning of 461 buildings within the flood zone.

"We spend a lot of time educating the community about climate change and predicting sea level rises, while also implementing and managing the projects," says Williams. "We are monitoring rainfall, water levels and the state of the natural defences. We are now moving to the next stage of the project, which is looking at what this means for local people and their best options. As civil engineering it is a very different challenge: it is data gathering, monitoring and informing."

YGC is involved in several other flood alleviation schemes, such as in the village of Llanberis, where an old masonry arch bridge over the local river was causing flooding. This was replaced and other flood defences were introduced in a £1.3 million project.

"Flooding and coastal erosion caused by climate change are two of the biggest natural challenges that will affect the safety and sustainability of communities across Wales over the coming century," says Huw Williams, head of YGC. "A key change is the priority and urgency about the timeframe of the decision-making process."

Environmental monitoring
for the digital age

QUALIS FLOW
Location: London, UK
www.qualisflow.com

Qualis Flow was born out of the frustrations of a civil engineer and an environmental scientist, both passionate about building a sustainable future. Brittany Harris and Jade Cohen met in New York in 2016, volunteering their skills to tackle the United Nations Sustainable Development Goals. On returning to the UK they began to pick apart their own project work-flows, and uncovered a wealth of opportunities to not only improve productivity and profitability in construction, but also empower teams to build more sustainably. They set up Qualis Flow – also known as Qflow – in 2018.

"In essence, Qflow is an environmental risk-management platform for construction," says Brittany Harris, co-founder and CEO. "We automate the collection, monitoring and analysis of environmental data from construction sites: from noise and air-quality data, to waste and materials transactions. Through tracking this data we identify when construction teams are non-compliant with regulations and notify the relevant people so they can take action straight away, preventing errors from becoming risks. We have carefully designed Qflow to be as simple and light touch as possible. We know teams don't want to be drowned in data, so we get the AI to do the heavy lifting in the background, and only provide the site teams with the information they need to get the job done."

But Qflow isn't just about streamlining data collection and reporting for construction. Its long-term vision is to learn from this data; providing projects with transparency and accountability throughout the build process, and ensuring that lessons learned from one site can be easily transferred to others. "In this way, we can begin to understand not only how to build with a more positive environmental impact, but also where the opportunities are to improve the sustainability of our designs," says Harris. "We can specify which projects reduce waste, which methods cause the least disruption, and source more sustainable material."

The company's platform has already been adopted by the likes of Canary Wharf and other tier-one contractors. Over time, Qflow hopes to expand from the UK into Europe and the US, building on its current portfolio of offerings. "We want to explore other environmental factors and carbon accounting," says Harris, "seeking new ways to help engineers build a sustainable future."

Flipping the energy switch

NEUERENERGY
Location: London, UK and Hoofddorp, Netherlands
www.neuerenergy.com

Switching to renewable sources of energy to reduce a company's carbon footprint is often painted as a straightforward undertaking, but it's actually a far more complex procedure than many might think. NeuerEnergy was founded in 2019 to help simplify this process, working to the benefit of both energy suppliers and energy users to help customers reduce their CO_2 footprint.

Using the latest information provided by data scientists, NeuerEnergy's platform allows customers to explore their options and comprehend costs and benefits right across their supply chain. It provides users with a simple-to-understand cloud-based dashboard that helps them keep track of their CO_2 levels and to perform cost-benefit analysis according to criteria that is tailored to individual business needs.

"Complexity and a multitude of external stakeholders are the main reasons for the slow uptake on renewable energy," says Muhammad Malik, NeuerEnergy's founder. "For purchasers it's currently a very fragmented journey that we are seeking to harmonise by giving the best possible advice on what to purchase and how to purchase it with the least amount of risk and highest confidence. For, suppliers we are enabling them to generate the equivalent of fossil fuel returns from renewable energy by aligning them with the other side of the chain. We streamline the risk management and ensure that the workflow and the financial and procurement aspects are harmonised in a clear way for all involved – that includes the purchaser, the producer and those in the middle, such as the national grid and regulatory bodies. We bring that entire ecosystem into a single workflow to automate the process."

Located in London and the Netherlands, NeuerEnergy is currently focusing its operations in Europe, working with global brands that have a significant presence on the continent. A core element of its offering is transparency, and it is dedicated to making it as easy as possible for companies to navigate the complex world of Corporate Power Purchase Agreements – long-term agreements that enable corporations to contribute directly to the development of renewable assets – and to understand the various costs and trade-offs involved in using green energy. NeuerEnergy will help a company offset its carbon footprint, find new suppliers or even construct its own renewable energy sources if it has sufficient space for a solar plant or wind turbines.

"We have developed end-to-end workflows and frameworks that are very persona-related," says Malik. "This means that the different people in a company – the chief finance officer, chief legal officer, corporate social responsibility officer and chief executive – can see the indicators that are most relevant and important to themselves."

NeuerEnergy collates intelligence through its own data engineering team, which processes raw data to do the necessary modelling, but this is supported by collaborations with other data organisations, whose information NeuerEnergy can draw on. This intelligence is used to help energy suppliers find customers and also manage the maintenance of their generators. "We work with a myriad of regional and global generators, and enable a seamless matching of supply and demand," says Malik.

This isn't just for big companies. The easily accessible process allows smaller companies with lower power needs to be automatically aggregated alongside larger companies or other similarly sized companies to reduce costs. "It's not just for large corporations," says Malik. "Our aggregation model can work with a small business or even an individual."

An amphibious wonder

WATERMASTER
Location: Finland
www.watermaster.fi

Finland is the land of 200,000 lakes but, until 30 years ago, there was no machine suitable for maintaining these vast areas of inland water. That's why the Finnish company Aquamec created the Watermaster, a specialist amphibious multipurpose dredger that can handle the numerous jobs needed to maintain shallow-water environments such as rivers, lakes, canals and coastal regions. It has since been used in projects in more than 70 countries as well as many of Finland's numerous lakes.

"The key element for global sustainability is a healthy and functional environment, and clean, healthy and functional waters and waterways play a huge role in this," says Aquamec's Senior Vice President Janne Heinonen. "Watermaster aims to help fixing these problems worldwide with innovative, proven and long-lasting solutions that enable doing more with less."

Since its launch, the Watermaster had been in constant development so it can continue to keep waterways in good condition, prevent floods, clean urban canals and remove invasive water vegetation. "Excavators are land-based machines and conventional dredgers can only operate in deeper waters," explains Heinonen.

"There wasn't any machine which could have properly handled the work in the shallow waters. The Watermaster can safely, efficiently and – in an environmentally friendly way – handle work in the whole shallow water area from dry ground to a water depth of six metres."

This attitude to the equipment reflects the company's overall approach to sustainability. The Watermaster is mobile and versatile, capable of doing many different jobs including dredging, excavating, raking and pile driving, which means operators only need a single machine to do their work.

"Cleaner, safer and better functioning waters with fewer machines and costs – this is the most sustainable way of solving shallow water problems," says Heinonen. "It's why we develop and service our products. We create new innovations that can be attached to older Watermasters to ensure they have a long lifespan. By buying a Watermaster, the customer gets much more than a machine – the customer gets constant feedback, support and new ideas to help to grow their own business in this demanding field of engineering. We help and educate our customers to operate our machinery in the most efficient and environmentally friendly way, to keep their machinery in good condition by maintaining them."

The benefits of
thinking long-term

BARHALE
Location: UK
www.barhale.co.uk

Many companies want to embrace a sustainable ethos only to discover that doing so requires a complete shift in perspective. At Barhale, sustainable engineering defines much of what the company has been doing for decades.

The engineering firm was founded in 1980 by Dennis Curran, a tunnelling specialist. "Dennis had seen some questionable practices and was determined to create a firm with different values," says director James Haddon. "Today we turn over around £200 million a year but we are still a family company and that means we can think about sustainability as we are not driven by short-term targets."

As part of Anglian Water's Integrated Operational Solutions Alliance, Barhale is currently concluding connection points for an initial programme of solar arrays at several Anglian Water sites in East Anglia. The new feeds to existing circuits help to provide clean, renewable energy for treatment works, with any surplus exported to the grid. These align with three of the UN's Sustainable Development Goals: clean water and sanitation; affordable clean energy; and climate action. However, while adding solar to the

power mix brings great benefits, it requires considerable engineering expertise. This includes trying to connect sites that were built decades ago and have their own unique constraints when it comes to incorporating new current generated from the solar arrays. While Anglian Water has established and maintained an industry-leading commitment to carbon reduction and sustainability, Barhale shares alliance responsibility for the design of the solar array from a safety perspective, ensuring that the design meets its client's standards.

The largest site, Grafham Water in Cambridgeshire, is almost as large as the rest of the programme put together and is one of the biggest privately funded solar projects to be completed in the UK, having been achieved with no government subsidies. Barhale was able to complete the connection work in line with the programme due to the expertise that comes from the way it has operated since it was founded.

"We aim to have a permanent full-time workforce," says Haddon, "which isn't typical of contractors of our size as there is much more to organise, but it means we can be in control and maintain our high standards. It's another example of our sustainable thinking."

Sustaining growth, sustainably

DELTA-SIMONS ENVIRONMENTAL CONSULTANTS
Location: Lincoln, UK
www.deltasimons.com

When Alex Ferguson completed his geology degree in the early 1990s, 90 per cent of his class went straight into the oil industry. Ferguson already knew that wasn't for him. "I was sure I wanted to be more focused on the environment and Delta-Simons is full of people who think like that," says Ferguson, who is now Managing Director. "We don't have a department of sustainability, we are a business that operates sustainably and we help others to do the same. It's all about facilitating a sustainable outcome, whether that is a new development, enhancing an existing one, making a difference to the social wellbeing of the people who work there or ensuring the financial sustainability of the investor or developer to make sure the cycle continues."

Delta-Simons was formed in 1992, originally concentrating on providing environmental assessments for the real estate sector. While that remains a core part of the company's offering, Delta-Simons' services have steadily expanded to include a fuller range of environmental services. "We do flood risk, air quality, ecology and arboriculture, bio-diversity enhancement, health and safety, due diligence for mergers and acquisitions, waste

management and wellbeing," says Ferguson. "We work with our clients to ensure they have the technical and commercial advice to do their business better."

These clients include multinationals, who draw on the expertise provided by Delta-Simons to ensure they meet their own sustainability criteria. The company is based in Lincoln but has around 150 staff across ten offices in the UK and Ireland. It is also a founding member of the Inogen Environmental Alliance, a group of 67 associate companies facilitating global service coverage; Delta-Simons has a place on the board.

For Delta-Simons, the broad approach of the UN's Sustainability Development Goals act as a helpful guide for companies who are wondering where to place their focus. "You have to pick the goals that you are best placed to tackle," says Ferguson. "The SDGs are a great way to break up the complexity of sustainability. They provide a framework against which we can discuss the issues and ask a business where they believe they can make the difference. It's about focusing and having a vision, so we ask which of these 17 goals can you really make a difference with?"

At the forefront of flood defence

VOLKERSTEVIN
Location: UK
www.volkerstevin.co.uk

In the low-lying Netherlands, issues surrounding water and flood defence have formed part of the national conversation for centuries. That's what makes VolkerStevin, the UK arm of Dutch engineering and construction firm VolkerWessels, so well placed to tackle some of the similar challenges now facing the UK.

Around 50 per cent of VolkerStevin's UK turnover comes from the water and environment sectors, including long-term contracts with the Environment Agency and water utility firms. "If your feet get wet, we step in," says Terry O'Connor, Sector Director.

O'Connor's plan is to use the timescale offered by these contracts to introduce new initiatives and then share them with the wider UK water sector. "These long-term frameworks mean we don't need to look at things in such a transactional way and can be a bit more forward thinking in how we approach things," he says. "I want us to be even more collaborative and to find better solutions that take time to develop.

"We are committed to sharing best practice and supporting a number of initiatives that should make a difference in carbon reduction, chemical usage and smarter technology," he continues. "Our approach will be that we can share something that we have tested and started to implement, as we will already be moving on to the next thing."

VolkerStevin is committed to finding sustainable solutions to issues around flooding. Technology will also play a major role; artificial intelligence is opening the door to unparalleled opportunities, and will help find solutions to problems well ahead of construction.

"Civil engineers usually want to talk about concrete, but for me this is about designing better and using technology to get a smarter solution, while reducing carbon and energy usage," says O'Connor. "One of the key issues on our agenda is that the problems around water caused by climate change, which we used to see happening in distant lands on our TV, are now happening on our doorstep. As a result, a new generation of engineers must lead the way in thought leadership. We need to understand what we are building, why we are building it and how to make it more resilient for the future."

Maximising efficiency, eliminating waste

RICARDO
Location: Shoreham-by-Sea, West Sussex
www.ricardo.com

Although his career as an engineer began more than 100 years ago, Sir Harry Ricardo always understood the importance of sustainability during his influential work designing and improving car, plane and tank engines. Indeed, when he formed Ricardo Ltd back in 1915, Sir Harry's mission was to "maximise efficiency and eliminate waste". This remains central to the company's approach today, which helps to align it fully with the UN's Sustainable Development Goals (SDGs).

Ricardo has three divisions – automotive engineering, rail, and energy and environment – with the latter best placed to support sustainability. "We work with national governments, international donors who are funding climate-change work, subnational organisations such as local authorities, and the private sector," says Trevor Glue, Director of Digital Services and Marketing. "We support organisations across the climate change and sustainability agenda. We use our extensive experience and technical expertise to support engineers to deliver the environmental-led components of their projects, like our work supporting water infrastructure in Jersey (pictured, right)."

The work Ricardo undertakes invariably overlaps with the United Nations' SDGs, particularly SDG 11 (sustainable cities and communities), SDG 12 (responsible consumption and production), SDG 13 (climate action) and SDG 14 (life below water). The company covers a wide range of environmental services to support civil engineers, helping them to innovate and implement energy projects. These include the development of sustainability strategies and the support and implementation of life-cycle assessments, Environmental Product Declarations (EPDs) and Environmental Impact Assessments (EIAs).

"We provide policy through to practice," says Glue. "We do an audit and then provide a company with guidance about the technology available to reduce their impact. A good starting point is to measure what you are doing, so you can understand the full environmental impact of your business or project; and

from there develop a road map to improve performance using a robust science-based approach.

Ricardo boasts a global and diverse range of clients and projects in partnership with Mott McDonald. "Our scientists and engineers can highlight practical solutions and develop long-term strategy," says Glue. "For example, Ricardo consultants are currently working with UK Water to support the water industry in developing a strategy for achieving zero carbon by 2030. Together we are developing a detailed plan of measures to achieve this target."

High ground on the high road

EUROVIA UK
Location: UK
www.eurovia.co.uk

How do you manage over 50,000km of the UK's highway networks sustainably? This question is answered by Eurovia UK, a business that delivers infrastructure services for local and strategic authorities across the country.

In 2019, emissions from road transport made up around a fifth of the UK's greenhouse gas emissions. "The future is in our hands," says Chief Executive Scott Wardrop. "For us, it's about how we as individuals can take responsibility and ownership for tackling climate change, responding as local teams, making a difference in communities and ensuring we have a green recovery for our country post-COVID."

In line with the UK government's legally binding target of having net zero emissions by 2050, Eurovia UK's eco-goals are being implemented in all aspects of the business. Already, Eurovia UK has been making a difference by creating bio-diverse, greener roads, improving air quality and mitigating impacts of climate change. Increasingly, roads are being built using recycled materials and the business now uses low-carbon asphalt as standard.

This ethos is evidenced by the company's highway service business, Ringway, which works in communities across the UK where, in 2020, widespread flooding hit large areas. "We're at the pointy end of climate change," explains Maggie Hall, Eurovia UK's Sustainability Advisor. "It impacts our people and what we do on a daily basis. We're cleaning gullies and keeping roads open, so the social and economic elements of sustainability are a driving force for us and the decisions we make as a business."

To help further its ambitious goals, Eurovia UK's consultancy arm; Jean Lefebvre UK (JLUK), is putting its technical and R&D expertise to good use. JLUK's laboratory is developing new, sustainable materials to be used on UK roads, tapping into a network of global practitioners to bring best practice to local highways.

One of the company's key assets is its people – and no single organisation can solve this in isolation. Eurovia UK engages with leading organisations such as ICE, ADEPT (Association of Directors of Environment, Economy, Planning & Transport) and CIHT (Chartered Institution of Highways & Transportation), and teams up with highway authorities and other sectors to work on practical ways to collectively lower carbon emissions. "Working with others is a powerful tool for finding and sharing solutions," says Hall. "Collaboration is key."

Beyond blasting

EPC-UK
Location: Alfreton, Derbyshire
www.epc-groupe.co.uk

"When explosives are being used on civil engineering projects it can often raise concerns, both socially and environmentally," explains Mark Jones, head of the Civil Solutions team at EPC-UK. "It's why such schemes need to be approached with sensitivity."

As the UK's foremost provider of commercial explosives and blasting services to the construction industry, EPC-UK has more than 115 years of experience in the construction, quarrying, offshore and civil engineering sectors. As such, the company is well versed in working on developments where it is paramount to consider sustainability, the local community and the surrounding environment. "All of these factors should be high on the agenda when planning and designing a blast and need to be taken into consideration from the outset of every project," says Jones.

One such development was an eco-friendly holiday park in a reclaimed quarry. It involved blasting a series of trenches for the park's drainage system, together with excavation work to create settlement ponds and a recreational lake – all of which had to be undertaken while remaining mindful of the project's environmental aspects. Another project involved stabilising

a rock face at Electric Mountain, a power station in North Wales, while ensuring that surrounding structures were kept safe.

"Both projects saw us working in very close proximity to buildings and public rights of way and as such, strict vibration criteria were imposed by local authorities and contractors," says Jones. "We have vast experience of working under these conditions and understand wholeheartedly the concerns of local communities with regards to drilling and blasting."

This is why EPC-UK is continually investing in the very latest technology. "We work hard to develop new techniques that enable us to carry out these necessary civil projects with the utmost precision," says Jones. "By accurately designing each blast using state-of-the-art software, no damage was caused, and we were able to keep well within the required vibration limits. Having worked closely with various contractors, clients, local communities and local authorities over many years, we have developed the expertise and knowledge to successfully deploy our drilling and blasting services while respecting the sustainable, environmental and social aspirations of every project on which we work."

Carbon-conscious engineering

CURTINS
Location: UK and Ireland
www.curtins.com

Be brave. That's the crucial final step of the seven-stage pathway to sustainability created by Stephen Beggs, the Technical Director of UK engineering consultancy Curtins, who has drafted the firm's carbon-reduction strategy. "It isn't easy to have that conversation with a client who has always done everything in concrete and steel," he says. "You need to be brave and you need to be brave in your own thinking about materials. Could we use a raft instead of piles? Can we upskill our knowledge of new materials? Do we need to take more risks?"

Beggs composed the strategy after Curtins added its name to the UK Structural Engineers Declare Climate And Biodiversity Emergency charter. He realised that if Curtins was to make a difference it was essential for staff to comprehend what was being discussed so he created a presentation for all 400 of them. "We showed them what a tonne of carbon looks like – if you fill two water butts with petrol and set fire to them they release about a tonne of carbon," he says. "We told them that a reasonably responsible personal lifestyle uses ten tonnes of carbon a year but their professional footprint making new buildings in steel and concrete is 1,000 tonnes. You could immediately see the reaction in the room."

After establishing awareness, the second stage is to measure the carbon in everything the firm makes and then include that information in drawings. That leads to the fourth stage, talking to clients. Stage five is to establish a hierarchy of alternatives – from scrapping the project altogether down to a last resort of offsetting – while stage six is to simply reduce overall use of concrete. And once that's all done? Be brave.

Young engineers are desperate to achieve sustainability and Beggs wants everybody at Curtins to start talking in the currency of carbon as well as sterling. "This is Curtins' 60th year and we have marked the anniversary by doing something very special and that's getting everybody on to the same page," says Beggs. "That means our structural and civil engineers, geo-environmental team and transport planners but it also includes the administration and support staff. Now they all realise what we are talking about and can join the debate with clients."

On point for the green revolution

CONNECTED KERB
Location: London, UK
www.connectedkerb.com

Electric vehicles offer a sustainable future for smart cities, but at the moment the uptake is being hindered by a lack of suitable charging points. That's what Connected Kerb set out to address when it was formed in 2017. The company has developed a new approach to electric vehicle charging infrastructure, creating an innovative and flexible concept that won the Mayor of London's Award for Civic Innovation in 2018.

"There is a very clear case for driving electric vehicles, both environmentally and economically," says Chris Pateman-Jones, CEO of a company that prides itself on being disruptive, visionary and bold. "But people need reliable and accessible infrastructure for that to happen."

The capabilities of Connected Kerb's solution reach far beyond that of traditional charging. Instead, the system acts as an integrated smart cities platform through its connection to both power and data. It is wireless charging enabled, can host Wi-Fi, 5G and various "Internet of Things" technologies, such as air-quality sensors. Crucially, the system is flexible to upgrades and new technologies, minimising any risk of obsolescence.

Reliability and adaptability are essential if every household in the UK is to have access to a charging point, so Connected Kerb's units are designed in a way that allows outdated or damaged parts to be replaced or updated with ease. The sensitive technology is housed below-ground and high volumes of recycled materials are also used, including recycled rubber and plastic. The end result is a reliable, sustainable and long-lasting solution to a problem that is delaying the much-needed mass take-up of electric vehicles.

The company wants its model to benefit all sections of society, not just those with access to private driveways and off-road parking. "There's a danger of creating a two-tier society when it comes to this technology," says Pateman-Jones. "We calculate that more than 60 per cent of the country does not have access to driveways where they could charge. We need charging points for these people, as they are more likely to be impacted by poor air quality and new penalties in cities for driving petrol vehicles." The company intends to install 3,000 charging points by the end of the 2020-2021 financial year, the first step on its target of becoming a global leader in sustainable infrastructure.

Roads to transition in sustainable transport

KEVIN MCSHANE LTD
Location: Belfast, UK
www.kevinmcshane.co.uk

"I'm not the sort of person who sits back waiting for others to do things," says Karen McShane, founder and Managing Director of Belfast-based transport and civil engineering consultancy Kevin McShane Ltd. Karen founded the company in 2015 after leaving the multinational she'd worked at for 25 years. She wanted to go it alone in business; and to transition to a woman.

"Over the past five years I have been able to take on two employees each year as we slowly expand the business," she says. "But we are not profit driven. Our ethos is to do a good job and make a difference to people's lives."

The UN Sustainable Development Goals are central to the company's approach and McShane is a staunch advocate of sustainable transport solutions. To that end, she has served on several professional bodies with a view to influencing policy. McShane encouraged the redesign of the central Belfast Royal Exchange development and undertook the removal of its 1,500-car basement parking provision. This drastically reduced the cost for the developer and meant historic facades and buildings could be retained. "We are now working actively with car club providers

to bring them to Belfast and provide people with flexibility," she says. "It's about changing the attitude of people towards travel and educating them about what we are doing."

The company has worked with the Royal Ulster Agricultural Society to provide transport solutions for their annual show, and is helping redevelop the RUAS's old site into a medical centre. "We have been looking at the best and most sustainable use of that site in terms of development and transport," she says. "We have brought in care homes and put in some car club spaces and pedestrian links to the nearest station."

McShane highlights the importance of education to the future of sustainable engineering, and her company currently employs two apprentices while encouraging staff to attend professional courses whenever necessary. Gender equality is another important issue for McShane. "I took a conscious decision not to hide in the background and I am quite vocal about who I am and what I believe in, while also being professional," she says. "I am first an engineer, then I am a transwoman."

A sustained defence of the nation

BABCOCK INTERNATIONAL GROUP
Location: London, UK
www.babcockinternational.com

Babcock International Group is a leading provider of complex engineering services which support national defence, save lives and protect communities. Focusing on three highly regulated markets – defence, emergency services and civil nuclear – Babcock uses technology to deliver vital services and manage complex assets in the UK and internationally.

"As a global, technology-led business we have a real opportunity to make a difference in the work we do and deliver, but we have to think longer term," says Dr Jon Hall, Babcock's Technology Managing Director. "Our people are working on some of the most complex engineering challenges that industry is facing today. Whether it's transitioning our customers to greener fleets, finding ways to reduce carbon emissions in their buildings, or increasing energy efficiency to prolong the life of an asset, then technology is a fundamental enabler."

For Dr Hall, Babcock's approach to sustainability is an integral part of its business, wherever it operates and wherever its people are. "This isn't just about setting a sustainability agenda," he says. "With over 34,000 people in our business we have to make it an agenda for them too and that means making STEM (science, technology, engineering and maths) a priority. By empowering our graduates and apprentices to get involved in our technology and innovation programmes we know we're building a legacy that will last. For instance, in the run up to World Environment Day this year, our graduates ran a campaign on how we can all develop a 'sustainable mindset'. They also harnessed their passion for technology and sustainability to reach the final of a national UK sustainability challenge to help the Ministry of Defence to reduce its carbon footprint through the use of technology."

Babcock's graduate trainees created a digital twin solution for a building which would be fitted with sensors, and the energy usage recorded would be analysed through artificial intelligence. This would then prompt what necessary energy improvements could be made. "Protecting our environment is a global challenge and we should all be asking ourselves what we can do to cut our own carbon footprint," says Josh Spencer, a civil engineering graduate. "Industry may be a big emitter of CO_2 but we all need to develop a sustainable mind-set. We're doing that at Babcock."

URBAN GROWTH

ABOVE

Delegates gather at the
UN–Habitat World Urban
Forum 9 in Malaysia, 2018

Home advantage

A housing project in war-torn Yemen shows the reach of UN-Habitat's vision to provide a better quality of life for all in an urbanising world

When fighting broke out in Yemen's West Coast Districts at the end of 2017, most of the people living in the town of Qataba fled and their houses were destroyed.

Mariam Saleh Mahdi and her family remained as they could not afford the transport, but their house was hit by shells and Mariam's sister was injured and died a few days later. The family walked to a nearby village and stayed for four months. When they returned in May 2018 they found their home had been completely destroyed.

"The walls and the roof fell off and the house had no doors or windows," says Mariam, who was looking after her sister's children. "The roof could not protect us from the rain anymore and the house was not safe to live in. I felt helpless, I had no idea how to rebuild the house on my own and with no source of income."

Mariam says that when the UN-Habitat team came to the village for the first time and started collecting data, she did not believe they would help. So she was surprised when she heard that her house would be rebuilt and that she would be paid to repair her own home as part of a cash-for-work initiative. She uses the cash to buy food for the family and plans to save for a television powered by solar energy.

REBUILDING LIVES

Mariam's house was rehabilitated as part of UN-Habitat's project "responding to the immediate needs of shelter, water, sanitation and hygiene in Al-Hudaydah Governorate, Yemen", funded by the government of Japan. The project aims to mitigate the impact of conflict on the internally displaced and vulnerable communities in Al-Hudaydah Governorate, with a focus on women, children, elderly, and disabled persons. It provides job opportunities for community members who are the main bread winners for their families by engaging them in the repair and rehabilitation works of their homes.

"UN-Habitat has enabled the rehabilitation of the entire house and provided us with electricity, lighting and a ceiling fan, all operating via solar power," says Mariam. "Our home has become safe to live in. When I see my new house, the latrine, the water tanks and the electricity, I feel overwhelmed with happiness. This helps me overcome the bitterness I went through because of war and displacement."

UN-Habitat was founded in 1978 as the United Nations Human Settlements Programme. Based in Nairobi, Kenya, it is mandated by the United Nations General Assembly to promote socially and environmentally sustainable towns and cities with the goal of providing adequate shelter for all.

UN-Habitat's vision of "a better quality of life for all in an urbanising world" is bold and ambitious. It works with partners to build inclusive, safe, resilient and sustainable cities and communities, and promotes urbanisation as a positive transformative force for people and communities, reducing inequality, discrimination and poverty.

UN-Habitat works in more than 90 countries to promote transformative change in cities and human settlements through knowledge, policy advice, technical assistance and collaborative action. Its mission can be summarised in four categories: think, do, share, partner.

By "think", UN-Habitat undertakes research and capacity building, sets standards, shares good practice, monitors global progress and formulates policies related to sustainable cities and human settlements. By "do", its operational work involves technical assistance and crisis response. The "share" side of its mission involves advocacy, communication and outreach; while "partner" involves collaboration with governments, UN agencies, civil society organisations, academic institutions and the private sector to achieve enduring results in addressing the challenges of urbanisation.

Supporting sustainable urban growth

UNOPS is working hard to help its partners deliver SDG 11 – to make cities inclusive, safe, resilient and sustainable

More than half the world's population lives in cities today. By 2030, according to a report in the journal *Science*, it is projected that six in ten people will be urban dwellers. In Argentina, one of the most urbanised countries in the world, around 91 per cent of people live in cities, while the metropolitan region of Buenos Aires alone is home to nearly half the country's total population.

Urban migration and continuous population growth have caused a housing deficit across Argentina. As a result, the country faces urban sprawl, overcrowding, a shortage of adequate housing and a lack of access to basic services. Informal settlements often spring up, posing health and security risks. Today, nearly 3.5 million households – more than 10 million Argentines – face a housing problem.

Infrastructure development in cities often fails to consider the interconnections between individual systems and sectors. Cities may lack knowledge on appropriate solutions to their most pressing needs, and a short-term bias in investments prevents understanding of the scale of losses in the future. However, sustainable development across the economic, social and environmental layers of society requires a long-term, evidence-based approach.

The social housing and infrastructure project – implemented by UNOPS in partnership with the Ministry of Public Works, local municipalities and community-based organisations – tackles urban poverty and promotes social and economic inclusion by improving living conditions, connecting homes to basic services and enhancing public spaces in vulnerable areas.

"Cities are the places where people look for new opportunities to advance socially and economically," explains Giuseppe Mancinelli, UNOPS Deputy Director for Latin America and the Caribbean. "Improving housing conditions is not only about the achievement of the Sustainable Development Goal 11, to ensure access for all to adequate housing and basic services by 2030, it is also about human rights."

Despite posing numerous challenges, cities can become incubators for innovation and growth, and drivers of sustainable development. This requires risk-informed planning and management, with extensive cross-sectoral infrastructure planning to ensure access to basic services.

OPPORTUNITIES AND CHALLENGES

New urban infrastructure must be safe and affordable. The burden of unsustainable infrastructure falls disproportionately on the world's poorest and most vulnerable, who often settle on hazard-prone, fragile land in cities. They are a key stakeholder when planning, developing and upgrading infrastructure.

Using digital technology, transport systems can be designed to reduce road accidents, while public transportation and sanitation facilities can be expanded to provide safer and accessible options

"The burden of unsustainable infrastructure falls disproportionately on the world's poorest and most vulnerable"

for all. Air quality and municipal waste and wastewater management require specific attention in cities to reduce adverse health effects – without proper sanitation services, the urban future simply will not be possible.

In addition, the climate emergency poses a looming threat. The risk increases in rapidly urbanising regions due to lack of planning and the low coping capacity of some infrastructure systems. Around $18 billion is lost per year in low- and middle-income countries solely due to the failure of energy and transport infrastructure following a hazard event. As such events are expected to become more frequent and severe, it is critical to ensure risk-management and finance mechanisms are in place to ensure rapid response and recovery, and to mainstream resilience into infrastructure decision-making to protect development gains and human lives.

The integral social infrastructure programme in Argentina is expected to benefit around 50,000 people, improving and building hundreds of homes, connecting around 2,500 homes to water networks, and around 1,400 homes to sewage networks. UNOPS works closely with local partners to promote employment opportunities to residents, ensure safe working conditions, advise on transparent administration of funds and supervise the quality of construction to ensure it is resilient and sustainable in the long-term.

Participants are also encouraged to take part in workshops on health and safety, safe use and management of water, waste management, nutrition and community integration.

The programme has also promoted alliances and partnerships with national universities and other UN agencies (such as UNAIDS and UN Women) to further build capacity and increase impact in communities around the country.

Extending the life of existing structures

Society no longer assumes that "new is best" when it comes to building and infrastructure – a realisation that has significant implications for engineers

SDGs: 9, 11, 12

Discussions about sustainability in the construction sector often focus on three key aspects: the need to use less energy in construction processes and the operation of buildings and infrastructure; economic design and resource efficiency; and using materials with the least embodied carbon. For the construction of a bridge or dam at a new location, this is clearly the right thing to do. However, what about existing infrastructure and engineering structures?

Taking a simple view, the most sustainable approach to engineering is, wherever possible, to avoid any new construction and continue to use what we have. There is also an intermediate approach between these two extremes – to extend the life of existing structures.

Of course, engineers have done this for millennia – repairing or strengthening old structures; increasing the capacity of reservoirs and bridges; adapting old buildings to new uses. Nevertheless, since the early 20th century, there has been a growing feeling among engineers and, especially, developers and asset owners, that "new is best" and we should get rid of the old. This was perhaps understandable in an age when material and energy resources were assumed to be infinite, and the impact of construction on the environment was largely ignored or seen as the inevitable price of progress.

GROWING AWARENESS

In the 1960s, many "inconvenient truths" became apparent. The destruction of habitats was exposed in 1962 in Rachel Carson's book *Silent Spring*; the "Blueprint for survival", published in *The Ecologist* in 1972, drew attention to the growing scarcity of many material resources; and the 1973 oil crisis highlighted the finite supply of oil. More significantly for the general population, perhaps, was the growing realisation that our built environment is something to be cherished and "new is best" began increasingly to be challenged.

A landmark example was the Supreme Court's refusal to allow developers to demolish Grand Central Station in New York. In London, the Euston Arch was not so lucky; it was demolished in 1961 and there was a great backlash of public opinion. In 1972, the "Convention Concerning the Protection of the World Cultural and

The refurbishment of
Blackfriars Railway Bridge
represents an impressive
combining of old and new
materials and design

Natural Heritage" was adopted by the General Conference of UNESCO. Today, it is thankfully almost inconceivable that old buildings, bridges or other constructions would be demolished unless absolutely necessary.

Since the 1970s, civil and building engineers have developed many ways of dealing with existing structures, both those of heritage significance and the far greater number of ordinary structures. This involves a rigorous process of assessing the condition of what exists and evaluating what may be needed to extend a structure's life; any proposed interventions must respect the existing structure while allowing for different possible futures. For example, an old bridge might simply be preserved as a heritage artefact, and a new, nearby bridge constructed; or an existing bridge can be strengthened and refurbished to carry modern traffic loads.

BLACKFRIARS RAILWAY BRIDGE

The refurbishment and adaptation of Blackfriars Railway Bridge in the heart of London is a recent example. The original wrought bridge comprising five arches was designed by Thomas Cubitt and completed in 1886. By the early 21st century, the railway bridge and adjoining station required extensive works to ensure it was able to keep up with passenger demand. The structure was widened by nine metres and strengthened – first the east side, then the west – and the bridge remained open to rail traffic during the entire five years of the engineering works, except for six days over the Christmas holiday when the tracks were realigned. New and longer platforms were built on the bridge and span the entire River Thames; a new canopy, with 4,400 integrated solar panels covering an area of 6,000 square metres, generates 1.1 MW of energy – about half the station's electrical needs. And a new southern entrance gives passengers access from the south of the river for the first time.

"The final structure is an unusual combination of modern design and Victorian materials, which needed painstaking attention and assessment," says Elspeth Baecke, the Senior Engineer for the Blackfriars Bridge Project at Network Rail. "The delivery of the project was a challenge to all parties involved and is an impressive feat of engineering. The bridge and its roof are already iconic symbols of the new age of the railway."

Today, engineers are developing a more culturally sensitive approach to the future of our built environment. It is now widely recognised by civil engineers and, particularly, by the owners of built assets, that there can be both commercial and environmental benefits to extending the life of existing structures. Increasingly, engineers now realise the great benefits in designing and building new structures without a limited design life.

For example, a bridge can be designed so that all the parts likely to decay with time, such as steel reinforcement, can be easily inspected and replaced when necessary. Ultimately, this approach is going to herald a revolution in civil and building engineering and, to some extent, represents a return to the engineering of earlier days. Indeed, Brunel designed many of his railway viaducts with structural components made of timber that could be easily inspected and replaced; and much housing and a great many factory buildings bridges constructed in the 19th century have proved to be long-lasting and easily adaptable to today's needs.

Cultivating green cities

Increasing urbanisation can go hand in hand with the expansion of green spaces and eco-friendly infrastructure, helping city-dwellers connect with the natural world

SDG: 11

Despite the importance of cities in our societies and increasing urbanisation, our ability to design and manage sustainable urban systems is still limited. Global urbanisation has distanced humans from nature, causing resource consumption at unsustainable rates, environmental damage and hotspots of greenhouse gas emissions. In addition, urbanisation amplifies temperature-related health risks and flood hazards, with direct consequences on the well-being of billions of city dwellers worldwide.

Reconnecting cities, and their people, to nature has become a common objective in the sustainable development debate. The expansion of urban green spaces, such as parks, trees, vegetated roofs and sustainable drainage systems, helps with some of the problems associated with urbanisation. Plants can reduce extreme heat (by shading and evaporative cooling), improve air quality, decrease flood risk (by reducing impervious surfaces) and increase habitat connectivity, thus enhancing biodiversity. Access to green spaces can also improve people's health, stimulating physical activity, and reducing anxiety and other mental-health issues.

The positive effects of nature on human well-being have been attributed to "biophilia", a fundamental need of humans to connect with the natural environment. Architecture has been largely inspired by ecology and new concepts such as arcology, biophilic design, green urbanism and eco-cities are now gaining momentum, capitalising on the need to address the global sustainability challenge. The vertical forest in Milan is one of the most recent examples of architectural structures seeking new trade-offs between the built and natural environments, but several other examples of green buildings exist, from the hanging gardens of ancient Babylon to the Hundertwasserhaus in Vienna and the ACROS Fukuoka in Japan.

In addition to such building-scale efforts, we are also witnessing a call for systemic city-scale solutions. The experimental city of Arcosanti in Arizona and the recent manifestos by architects Stefano Boeri ("Urban forestry: a call to action") and David Chipperfield ("What is our role?") are clear examples of this vision. Policymakers and urban-planning authorities are also becoming aware of the need for green infrastructure as demonstrated by the London Environment Strategy and the declaration of London as the world's first "National Park City". Similar examples around the world include the Singapore Green Plan, the 20 Million Trees Program by the Australian Federal Government, and the Los Angeles County Safe Clean Water Program. In the light of these efforts, engineers should also reflect on Chipperfield's question: "How should we react professionally to the challenges of climate change and growing social inequality?"

New materials and smart technologies alone will not create smarter cities; we need to rethink our societal values and long-standing separation from nature. The New Urban Agenda adopted at the UN in 2017 encourages "nature-based innovations", as well as "robust science-policy interfaces in urban and territorial planning". Engineers should optimally include urban vegetation in their designes, and not just for aesthetic reasons. Traditional engineering (from structural design, to mechanics and hydraulics) must be blended with plant physiology, ecology, biogeochemistry and atmospheric sciences to quantify the costs and benefits of green infrastructure, for everything from buildings up to whole countries. Green spaces should be made accessible to all, for better social equality and to improve the living conditions of the most vulnerable communities.

This all requires close collaborations with experts from different professions and a shift of values – from wealth to health. If well "engineered", green infrastructure has the potential to shape healthier, more sustainable and more just cities.

Liveable neighbourhoods

Waltham Forest's "mini-Holland" scheme has created safer public spaces and helped reduce air pollution

SDG: 11

From the 1950s onwards, the urban fabric of cities has been substantially altered to accommodate the car. This was achieved by the construction of new routes for cars, and area-wide motor traffic management systems. It has resulted in cars dominating the urban public realm, including minor roads. Parked cars using valuable space is a major issue in many cities. With some notable exceptions, such as Houten in the Netherlands, new settlements have generally been planned with the primary user of public space being motor traffic.

This car dominance has many adverse effects. It constrains the movement of those walking and cycling within their local communities. It also results in poor air quality that is hazardous to health, and an environment that kills and seriously injures millions of people every year worldwide. Car-based transport systems are also prone to congestion because of their inherent inefficiency. The result is an energy-intensive transport system that produces a significant proportion of developed countries' carbon emissions. The current situation does not meet the aims of Sustainable Development Goal 11 to make cities and human settlements inclusive, safe, resilient and sustainable.

One solution is to re-engineer neighbourhoods in a way that makes them more "liveable" once again. The London Borough of Waltham Forest in north-east London has successfully created such liveable neighbourhoods, under a funding programme called "mini-Hollands". The project includes the introduction of protected space for cycle traffic on main roads. Cycling is efficient in space use, and poses less risk to other road users than driving does. The resulting space is attractive, comfortable and safe; all necessary pre-conditions for city dwellers to readily choose cycling. To be useful, networks for cycle traffic must be fully inter-connected and create route opportunities that are as direct as possible.

RE-ENGINEERING THE PUBLIC REALM

Importantly, the Waltham Forest project also limits the movement of motor traffic by selective filtering of traffic. A key feature of

creating sustainable communities is removing cars from areas that can then be used predominantly by the people living in the neighbourhood. While the concept is straightforward, it can sometimes be difficult to persuade a local population of the ultimate benefits of re-engineering the public realm in this way.

Footways were constructed that continue over side roads where they connect with main roads. These continuous footways offer protection and priority to pedestrians, and also cyclists where there is a cycleway present. Through clear and unambiguous design, they reinforce regulation, which affords pedestrians priority. Continuous footways have been widely used in Denmark.

Other features include walking and cycling routes that make it easier to travel between the different local centres; cycle hubs and cycle hangars to make storing cycles easy; and new and improved public spaces to improve air quality and make the borough a more attractive place to spend time. Longer-distance trips within the wider city are enabled by networks of buses and underground and overground railways. In addition, bus stops and bus stop locations were upgraded to make public transport more accessible and reliable.

Early indicators suggest that people in the area have increased their walking by 32 minutes a week and cycling by nine minutes a week. As a result of this and improved air quality, life expectancy has increased by seven months. There has also been an increase in 30 per cent in retail activity and a 17 per cent reduction in the commercial property vacancy rate. The number of people using the public realm for sociable activities has doubled.

A sustainable city is a city designed for the well-being of its people and the planet. The economic, social and environmental impacts and benefits need to be balanced, and to not store up problems for the future. The Waltham Forest liveable neighbourhood project provides an excellent example of how to achieve more sustainable communities through enhancements in mobility provision.

Re-zoning
the city

**Civil engineering has helped transform post-industrial cities
into areas that are more inclusive, safe, resilient and sustainable**

SDG: 11

Cities are continually evolving in response to economic, social and environmental drivers. Globalisation is accelerating this process and cities that are unable to respond may quickly lose their purpose and vitality. Well known examples of cities that have fallen victim to this change include Detroit, the "motor city" that was home to the US car industry, and Sheffield, the "steel city" responsible for the UK's stainless steel production.

City disruptors can be thought of as events, people or technologies that can rapidly change how a city functions. Technology-based disruptors (such as the internet and smartphones) and new service providers (such as Google, Apple, Amazon, Deliveroo, Netflix, Uber and Airbnb) are all changing how people live, work and play within cities. It is clear that many established city models such as large-scale retail are becoming redundant, creating the opportunity for cities to change to deliver the inclusive, safe, resilient and sustainable urban environments that people need.

In 2009, New York established the High Line, a 1.5-mile-long elevated linear park, formed from the spur of the former New York Central Railroad in Manhattan. The abandoned railway in the old Meatpacking district has been reinvigorated as a "living system" and planted with vegetation that would have grown up around the railway line once it fell into disuse. The High Line has become a major tourist attraction in New York, with 5 million visitors annually, and has resulted in wider regeneration along its route.

RADICAL THINKING

Medellin in Colombia, a city of 2.5 million people, is home to a visionary intervention, where EPM, the city utility company, developed public space parks called UVA (Unidades de Vida Articulada or "articulated life units") by re-zoning their infrastructure. The radical thinking here was for a utility company to take the city's water tanks and to transform them from gated, hard infrastructure to open, community spaces in a city where green space is at an absolute premium. The UVA parks' construction started in 2013 and, by September 2015, eight of them were finished and in operation.

Each park provided a computer room, advisory centre, classrooms, toilets, public realm, water fountains and lighting. To test the public's view of the parks, 100 people were surveyed by questionnaire in two parks across the weekdays and weekends. The perception of EPM as a company improved alongside security around the park and the general sense of community.

Local authority data also showed a reduction in violence around the parks. In addition, the surveys showed a degree of mismatch between what the public believe the parks should provide and what the UVA designers perceived. This indicates that constant evaluation and transformation of the parks is desirable to keep the parks relevant.

In particular, the green space was valued highly by the park users alongside the opportunity to have outdoor activities such as picnics, outdoor meetings and children's play in a safe environment. The UVA project demonstrates that thinking differently about large civil engineering infrastructure in a city and breaking away from the "business as usual" approach can transform the lives of city residents.

OPPOSITE

The High Line in New York reinvigorated an abandoned railway and now attracts some 5 million visitors a year

Joined-up thinking

Sustainability-assessed improvements to Stockholm's tram system are enhancing access for all

SDGs: 3, 10, 11

The Stockholm city tram system is expanding to link to Lidingö, a municipality to the east of the city, via Stockholm Royal Seaport, a new sustainable urban development. The first stage of this expansion is the accessibility upgrade of the Spårväg City tram line (which operates under the name of "Tram Line 7") and the tram stops in the city's South Djurgården district.

South Djurgården is the world's seventh-largest entertainment centre with more than 14 million visitors annually, and this figure is expected to rise; hence Tram Line 7 is an important connection for Stockholm's economy. Around 12,000 new homes and 35,000 new workplaces are under construction in Stockholm Royal Seaport, and the Spårväg City tram line will supply most of the public transport services.

The project integrated the civil engineering evidence-based assessment scheme CEEQUAL as a tool to provide guidance in every phase of the project, providing traceability in all aspects of sustainability. The focus shifted towards a wish for continuous

OPPOSITE AND ABOVE

The line upgrade ensures
that all passengers have
access to buses, trams
and platforms

improvements across the whole project (client, designer and contractor), which achieved a CEEQUAL construction rating of "Excellent" and was highly commended in the scheme's outstanding achievement awards.

The accessibility upgrade of the Tram Line 7 project was procured explicitly as a collaborative project between the two major stakeholders – the client, Trafikförvaltningen (the Transport Administration Stockholm County Council), and the contractor, Skanska. The two worked together to find solutions that are environmentally and economically responsible, as well as taking into account health and safety. Project members were committed to working towards common goals and to coming up with innovative solutions, and regular meetings were held between the project management groups of each organisation. Environmental and sustainability issues were prioritised in the project agenda through collaboration and the use of CEEQUAL.

SUSTAINABLE SOLUTIONS

The use of temporary tramway points has for many decades been a "forgotten technology" in Stockholm, one which the project revived. It enabled tram services to continue during construction to the benefit of both passengers and the construction team. Passengers were able to continue travelling by tram, avoiding the need for replacement bus services, which reduced the environmental impact of the project, reducing emissions as well as its cost.

In addition, the tramway runs parallel to the tree-lined avenues on Strandvägen, and there are trees in Djurgården that grow in close proximity to tram installations. This was taken into consideration and the team had a close dialogue with the Royal Djurgården Administration, which manages the land and trees in the area, to avoid damaging trees during construction and operation. Just one example of this is the measures taken to protect a specific lime tree on the platform at the Nordiska museet stop.

Accessibility was improved on Tram Line 7 to ensure that all passengers have access to and from buses, trams and platforms. Test platforms were built to assess the best design for accessibility features, and staff and representatives from various disability organisations tested these platforms.

A top priority of the project was to achieve a high level of third-party satisfaction (80 per cent) after the upgrade. A survey of perceived quality was carried out every month and local trains achieved 86 per cent (compared to the other modes of transport: metro achieved 82 per cent, rail 75 per cent, bus 74 per cent). For Tram Line 7, customer satisfaction scored 93 per cent, the highest level for any transport mode in the SL (Storstockholms Lokaltrafik) network.

Setting the standard in decontamination

The award-winning transformation of an old coking works near Chesterfield required imaginative remediation and restoration, with local community input

SDGs: 3, 8, 9, 11, 15

The former Avenue Coking Works at Wingerworth near Chesterfield in Derbyshire was one of the most contaminated sites in Europe. The plant opened in 1956 and produced smokeless solid fuel through the carbonisation of coal. It also produced town gas, for domestic use in Chesterfield, and generated electricity for its own use and the national grid.

Following its closure in 1992, the works lay disused until East Midlands Development Agency (and subsequently Homes England) commenced remediation of the 98-hectare site – the size of around 140 football pitches – in 1999. The project went on to win awards for the outstanding restoration of contaminated land.

Sustainability was at the heart of the scheme from the beginning. Starting out as a derelict industrial space, the client wanted the site to be returned to good use, for the benefit of the environment and the local community. The project needed to be as sustainable as possible, so good sustainability practice was incorporated into how the project was set up and implemented. It was decided to track this using CEEQUAL sustainability assessment.

The local community, regulators and other stakeholders were engaged proactively from the start. The outline masterplan was communicated via a series of public consultation roadshows, and the local community were invited to share their views on what they wanted the site to look like and what it would be used for when completed. Where appropriate, these ideas were incorporated into the masterplan design.

Around 70 hectares of public open space was created. The River Rother was realigned and restored to a cleaner and more "natural" state. High-quality sports pitches, new areas

LEFT AND OPPOSITE
The 98-hectare site is now home to significant new areas of wildlife habitat

of woodland and multi-user trails connecting to local footpath and cycle networks were added to this new space.

PROTECTING WILDLIFE

Following the closure of the coking plant, the site was colonised by a number of protected species, such as great crested newts, water voles, kingfishers, grass snakes, badgers and lapwings. Because of this, extensive wildlife surveys and exclusion works were conducted to allow construction work and remediation to take place. Habitats were created to provide suitable alternative areas for water voles and the translocation of some 8,000 newts and other species was arranged to allow the works to go ahead. The site now includes significant new areas of wildlife habitat, mainly due to the remediated River Rother.

During remediation activities, a water management plan was needed to separate and manage clean and dirty or contaminated water. Contaminated water was treated on site for disposal to sewers, and clean water was used on site where possible, for example in dust suppression.

The project aligned with the principles of the waste hierarchy by re-using on site all materials that could technically be treated. The landscape was designed to incorporate all the treated materials, with a minimum of new or reclaimed materials. Materials were remediated to be safe for re-use by numerous methods, including thermal desorption, ex-situ bioremediation, enhanced separation and screening, and concrete crushing and grading.

Apart from relatively small volumes of hazardous substances, such as asbestos and tar, which were sent for off-site disposal, the vast majority of material, following clean up, was re-used in appropriate locations across the site. This resulted in big savings in transport and associated carbon.

A collaborative line of thought

Major improvements to the West Coast Main Line through Staffordshire were carried out with sustainability and public-engagement at their core

SDG: 11

The Stafford Area Improvements Programme (SAIP) removed a major bottleneck on the West Coast Main Line railway, which runs from London to Scotland. Constructed by a team called the Staffordshire Alliance, the programme included a combination of 18 miles of line-speed improvements, the provision of 10km of new 100mph railway, including a flyover over the Main Line, 11 new bridges, four river diversions, major environmental mitigation works, and gas pipeline, road and footpath diversions. The project facilitated faster, more frequent and more reliable journeys between London, the Midlands, the North West and Scotland.

The programme was completed on budget and more than a year ahead of schedule. The Staffordshire Alliance team, with its highly collaborative approach to delivery, was recognised by the industry for being the first "pure alliance" for rail design and construction in the UK. It was also acclaimed for leading the way in innovation, and for demonstrating the viability of the alliance model. The team received several honours, including those for Collaborative Working and for Outstanding Large Project (of more than £20 million) at the UK Railway Industry Awards, and the best Major Project accolade at the Railway Industry Innovation Awards.

Network Rail registered the project for a CEEQUAL Whole Project Award to demonstrate how sustainability was embedded into its processes through an externally verified and rigorous assessment scheme. The Staffordshire Alliance team focussed on sustainability and used the CEEQUAL assessment process to drive sustainability in the design development. This approach continued through the construction phase, with CEEQUAL scoring incentivising the team to deliver the scheme in a sustainable way. The scheme has since won CEEQUAL's Outstanding Achievement Awards 2019 in the Water Environment and Resources category and was Highly Commended in the Ecology and Biodiversity category, with judges noting that it "achieved excellent outcomes way beyond what normal practice would require."

ENVIRONMENTAL ELEMENTS

The river diversions used the principle of natural channel design to ensure that the channels maximised their biodiversity value. In total, 830m of new channel was constructed, along with 237m of wetland habitat on both the River Meece and the River Sow. The design provided a net increase in channel length through a meandering channel design. The rivers were designed to be self-sustaining, being allowed to freely adjust, creating new forms and features as they do, thereby lowering the need for maintenance.

In river diversion locations, sections of the old channel were retained to create backwaters and a more diverse habitat. New wetland areas were created, suitable for rare invertebrates found on-site. Sustainable Drainage Systems (SuDS) were implemented to attenuate the water run-off from the new rail and road infrastructure – these have a built-in biodiversity benefit through planting and soft engineering.

The project also delivered benefits to the water environment. It included a re-rectioned hillside with small bridges designed to throttle flood waters, slowing flood flows into Stafford without causing significant damage or disturbance in the vicinity. This also helped save around £6 million that otherwise would have been spent on an alternative large viaduct.

Savings were made on energy and carbon, as well. Material use and waste was minimised, saving significant amounts of traffic, and the programme re-used 98 per cent of excavated materials on site, saving 124 daily 20-tonne lorry trips leaving site (including the returning empty lorries, that amounts to 27 movements an hour) for 10 months via a rural network.

The programme also aimed to establish a positive legacy for its neighbours. It established six replacement ponds near the site to mitigate the impact on local amphibian populations. These included creating a new community open space with an all-access dipping and viewing platform and boardwalk. And through a comprehensive consultation process, the local community helped influence the chosen rail, road and footpath alignments.

Building Britain's tallest bridge

Sustainability measures played an important role in the planning and construction of the Queensferry Crossing in Scotland

SDGs: 9, 11

Multiple ecological measures were included in the construction of the Queensferry Crossing

The Queensferry Crossing is a road bridge near Edinburgh, Scotland, which was built alongside the Forth Road Bridge. It opened in 2017 and safeguards one of the most vital connections in Scotland's transport network. It is the tallest bridge in the UK, and an iconic structure with a 2.7km span over the Firth of Forth. More than 23,000 miles of cabling suspends the bridge from three narrow and elegant towers.

As a dedicated public transport corridor, the existing Forth Road Bridge provides additional infrastructure capacity for sustainable forms of travel, including walking and cycling. The project achieved a CEEQUAL whole project rating of "Excellent", demonstrating the high level of sustainability performance.

Aesthetics were a major consideration in the design of the bridge. New roads were integrated into the surrounding landscape and screening was provided where required. Over 400,000 new trees, grown from the seeds of local trees, were planted. The new planting comprises large blocks of woodland, smaller areas of scrub woodland, hedgerows and individual trees. Two areas were planted before construction works started and so benefited from at least five years of growth before the new roads opened.

ECOLOGICAL MEASURES

The project aimed to minimise the impact on the local ecology and biodiversity. The Firth of Forth has a number of protected ecological sites, including Special Protection Areas, Special Area of Conservation and Sites of Special Scientific Interest (SSSI). Before any construction work started, a large number of ecological surveys were commissioned to ensure appropriate mitigation measures were designed into the scheme. These included construction-noise limits to minimise disturbance to birds in the marine environment. In addition, extensive mammal fencing and several mammal crossings were incorporated into the new roads.

An unavoidable impact from the project was the direct loss of land within St Margaret's Marsh SSSI, which is noted for two habitat features: transition saltmarsh (reedbed) and saltmarsh. In consultation with Scottish Natural Heritage, the remaining marsh is being restored. A management plan was developed and a steering group was established for oversight, including input from local community groups and other organisations.

To help the saltmarsh vegetation recover required hard engineering to improve the water balance on the marsh and provide increased inundation by sea water. This involved the construction of three sea-wall sluices to the Firth of Forth, in addition to five internal sluices between the compartments within the marsh.

It was important to minimise embodied carbon, which resulted in making maximum use of existing infrastructure, reducing the scale of new construction and applying best practice during design and construction. This included retaining the Forth Road Bridge as a public transport corridor and an improved active travel route. During construction, the contractor minimised carbon emissions by using a cut/fill strategy to avoid the requirement for materials to be transported off-site; using recycled concrete for fill and drainage; and having an on-site concrete batching plant to minimise transportation. During operation, maximum use was made of Intelligent Transport System (ITS) measures to improve traffic flow and minimise stop-start traffic conditions.

A range of communication methods also helped to keep local communities and wider stakeholders informed. A Contact and Education Centre was built for face-to-face engagement and also provided facilities for exhibition panels, presentations and extensive educational programmes.

A quiet revolution

The Thameslink Programme has succeeded in carrying out major structural work with minimal disruption to London's residents and rail users

SDG: 11

The Thameslink Programme was a £7 billion government-sponsored project in the south-east of England to upgrade and expand the Thameslink rail network. It allowed more people to travel on new and longer trains between a wider range of stations to the north and south of London. Platforms were lengthened, stations were remodelled, and new railway infrastructure and trains were added. This all provided a large increase in railway capacity through central London.

London Bridge station was the key component of the Thameslink Programme, as it was operating at maximum capacity. The renovation of the station was necessary to increase the number of Thameslink trains able to pass through the station and to meet the anticipated growth in passengers.

Sustainability was at the heart of the Thameslink Programme; it was important to deliver transport benefits to budget, to show value for money and to create an overall positive impact on the community and the environment. As part of this vision, the project's team decided to use the CEEQUAL methodology to help strike a balance between achieving high standards of environmental and social benefits, and delivering value for money.

BENEFICIAL MEASURES

A range of features included in the station's redevelopment resulted in total carbon savings of 18.9 per cent. Examples include the concourse, which was designed to allow in more natural light; the use of LED lighting; and "slow and go" escalators, which are predicted to save nearly 35 tonnes of carbon a year. Geothermal piles and a ground-source heat-pump were used for retail outlets, which will provide annual carbon

savings of 311 tonnes. In addition, reinforced steel was used, with 98 per cent recycled content, which has delivered a saving of 8,353 tonnes.

Major infrastructure programmes such as the Thameslink Programme inevitably generate noise and vibration, which can result in some of the most significant impacts on a local area. To deliver a sustainable project, the potential effects of noise and vibration were carefully considered during the planning process in an effort to reduce their impact.

The project's Noise and Vibration Policy set out the broad approach that was adopted. The underlying principle was to avoid significant adverse effects arising from either construction or the operation of the scheme, wherever and whenever reasonably practicable. Control measures were then developed in accordance with a defined mitigation hierarchy.

As a result, the project was highly successful in gaining the trust of local residents. This in turn delivered a distinct benefit to the project – consent applications were granted for extended working hours overnight and at weekends when necessary. This helped enable the project to keep on schedule. As an indicator of the success of this approach, the project was awarded the Noise Abatement Society's John Connell Silent Approach award.

The Thameslink Programme project demonstrated exemplary sustainability performance, and achieved the CEEQUAL Excellent Whole Project award with a rating of 97.3 per cent. The project also won two CEEQUAL Outstanding Achievement awards in 2019 for reducing the effects of construction on neighbours and protecting the historic environment. It was also highly commended for its energy and carbon savings.

Border connections

The UK's longest new railway in 100 years, connecting Edinburgh with the Scottish Borders, has delivered a positive, sustainable results

SDG: 8, 9, 11

In 1969, the Waverley Line, which passed through the Scottish Borders, was closed as part of a wider series of cuts across the UK rail network that followed two reports authored by Richard Beeching. In the years that followed, Midlothian and the Scottish Borders faced long-term social and economic decline following the collapse of the mining and textile industries. In 2006, the Scottish Parliament passed the Waverley Act Scotland, authorising the return of train services between Edinburgh and Tweedbank.

The resulting Borders Railway project involved re-establishing a 30-mile stretch of the former Waverley Line, closed for more than 40 years. This included the construction of seven new stations along the railway's path, offering enhanced connectivity to the rural communities on the route.

"Arguably Scotland's largest civil engineering scheme in recent years, the Borders Railway has focused on a sustainable legacy with the revitalisation of a significant number of structures previously left redundant following the lines closure in 1969," says Dan Whiteley, Head of Environment for BAM Nuttall, the project's principal contractor.

The project supports the Scottish government's strategy of ensuring that transport connections are strengthened and made more reliable to maximise opportunities for employment, business, tourism and leisure. The new line is an essential passenger link from the Scottish Borders, with a half-hourly peak-time service and journey times of less than one hour. The new railway provides a much-welcomed opportunity for economic growth in the area by improving accessibility for those keen to visit for tourism and leisure. It also improves the employment opportunities for residents previously constrained by the existing transport links, who are now able to travel easily into Edinburgh city centre.

LOCAL ENGAGEMENT

The project's recruitment strategy was designed to ensure that opportunities were provided to local businesses and personnel wherever possible. It had to compete for labour resources with other major schemes in the area, and sourcing local labour was challenging at times, but a commitment to local employment and engaging

The Borders Railway project has connected communities and created local employment opportunities

a local supply chain ensured that communities benefitted from the project, and many local residents were employed.

Community engagement played an integral role in the project's success. A dedicated community engagement team was established, through which the project team actively communicated with 26,000 line-side neighbours using community forums, drop-in centres, school visits and proactive social media campaigns. They responded to over 1,500 specific enquiries and reached more than 500,000 visitors via the project website.

A value engineering exercise was conducted to assess re-using existing structures compared to building new ones. The principal approach in its design was to re-use wherever possible the old infrastructure of the original railway, including structures, earthworks and tunnels. This made sense from both a cost and an environmental point of view. The original track alignment was also maintained where possible, helping to preserve much of the heritage of the railway.

Where new works were required, the project team developed designs that were cost effective and repeatable at various locations along the 30-mile route, including pre-cast bridges and standard design details. Concrete structures were preferred to steel to minimise both initial and long-term maintenance costs. In total, 95 bridges were regenerated and two iconic tunnels were resurrected to service the new route. An additional 42 new bridges were constructed and more than 1,500,000 tonnes of earthworks were removed to create the railway's path.

The project was commended by the Saltire Society in association with ICE for demonstrating an exceptional commitment to collaboration, coordination and communication with its stakeholders to deliver the longest new domestic railway in Britain for more than a century. The completed project has delivered great social, environmental and economic benefits to the area.

"This scheme brings together the engineering strengths of the BAM Group and created significant local employment opportunities," says Stephen Fox, former Chief Executive of BAM Nuttall. "In addition, it provided a much-needed rail link to those rural communities living in the Scottish Borders region."

The laws of the environment

CLYDE & CO
Location: London, UK and global
www.clydeco.com

The climate is changing. A recent scientific study showed that the number of potentially fatal humidity and heat events doubled between 1979 and 2017, and that they are increasing in both frequency and intensity.

Buildings and construction are major contributors to climate change – together they account for 39 per cent of energy-related CO_2 emissions. The construction sector is under growing pressure to decarbonise, with concrete and steel coming under particular scrutiny because of their high "embedded carbon", and to ensure that new buildings have lower energy intensity per square metre.

Five years ago Innovate UK published a report on adapting buildings to climate change. "Building designers have a professional duty to understand the potential implications of climate change, discuss them with clients, and act accordingly," it read. "Over time, it seems likely that liabilities will arise for building designers to take reasonable account of future climate change. As a consequence, building designers should at least inform clients about climate change risks, and record the outcome."

International law firm Clyde & Co has developed a climate risk practice which is headed by partner Nigel Brook. "The volume of data and knowledge about climate change has grown substantially since the Innovate UK report came out, and this duty of care can be regarded as firmly established," he says. "A failure to address climate change – whether the physical effects, or future changes in expectations and regulations – could give rise to a claim for damages."

Clyde & Co is curating various events and conversations to allow stakeholders to share current thinking in areas such as sustainable engineering and reducing carbon emissions. It conducted a study looking at how drones can be used to create a digital twin of a building, and then how to use that data to model how the building will adapt over time. Clyde & Co also looked at how smart contracts could automate allocation of defined risks, such as the impact of adverse weather on construction projects, based on objective data from an approved source.

"You have to think ahead," says Brook. "With climate change advancing and steps being taken to address it, the past is no longer a reliable guide to the future. The law will adapt too, and we are helping our clients to navigate their way through this fast-changing environment. The companies that thrive in the 2020s will be agile and adaptive."

Home truths for
a sustainable future

PROJECT ETOPIA
Location: Cambourne, Cambridgeshire, UK
www.projectetopia.com

The founder of Etopia, Joseph Daniels, is not your typical entrepreneur. Still only 28, he had a very difficult upbringing, bouncing from school to school with an alcoholic father and a mother who experienced mental health issues. Daniels had been homeless four times before he was out of his teens. It's this traumatic experience that has driven his desire to become an innovator in the construction industry. He wants to build better, cheaper, smarter, more energy-efficient, modular homes because they will save lives. And he hopes he can inspire one or two lost children like himself along the way.

"We're a construction-, energy- and intelligence-based company," he explains. "Etopia stands for economical environmental utopia. Utopia isn't achievable but it's an ideal and we want to strive for it through economics and the environment. We innovate technologies in the construction side, such as building materials. We then work on the generation and storage of energy, and we add intelligence. We ask how our homes can become better connected, how we can make every single piece of electronics in the home intelligent."

Daniels has devised a framework to inform this ambitious thinking. When looking at any new product or technology, Etopia considers it from both an economic and an environmental perspective. "With economics, you need scalability and you also need affordability – can it be built and can people afford it?" he observes. "The environmental factors that we take into consideration are sustainability and social acceptance – it has to help save the planet but people need to want it. We understand there is a matrix that must be achieved if we are going to take an idea forward."

Daniels' first success was the construction of a modular school extension in Brightlingsea, Colchester that employed cost-saving energy technology. The exterior shell was constructed in eight hours, and the entire building was completed in six weeks. Project Etopia has since been constructing a 47-home eco-project, Etopia Corby in Northamptonshire, consisting of modular houses that produce and store their own energy. The superstructures of these three-storey town houses were completed in 34 days.

Such success has attracted national attention, but Daniels' focus goes even further than this. His company's Seven In Seven programme has global ambitions. "I went through the UN's Sustainable Development Goals (SDGs) and picked out the seven key problems that can be solved with seven solutions on seven continents," he explains. "There are many SDGs but some aren't primary problems while there are a select few that, if we don't solve, the world will combust. So I found seven key problems – things like waste, food, water and overpopulation – that are shared by everybody and we are using our modular building system as an exemplar of how we can solve these issues on each continent."

For Daniels, it all goes back to personal experience and his determination to ensure people don't suffer the difficulties he had to overcome. "One time my mum had just come out of hospital and we ran out of food and electricity and the landlord wouldn't fix the boiler," he says. "These were terrible conditions. This wasn't the first world. When I decided to apply myself, my mind always went back to this moment and I began to think – why don't we just build stuff properly? Why can't that be available to everybody? I want to create something that will last longer and will sustain and improve people, and I want to inspire kids like me. That is something I feel passionately."

A long-term tunnel vision

GALL ZEIDLER CONSULTANTS
Location: Croydon, UK and worldwide
www.gzconsultants.com

When Kurt Zeidler, co-founder of tunnelling specialist Gall Zeidler Consultants, says that tunnels might have "an image problem", that isn't because of anything they are doing wrong, but simply a reference to their visibility. "People don't always appreciate what is happening underground," says Zeidler. "When you see a bridge, people know what it is doing. But tunnels present a huge opportunity for sustainability. The very principle is already helping to reduce environmental impact if we are doing things right."

Gall Zeidler is a worldwide leader in geotechnics, and tunnel engineering and construction, having worked on numerous international tunnelling projects. Zeidler believes a tunnel can go a long way towards delivering sustainable cities by freeing up space on the ground that can be used for residential developments and even the creation of new green spaces. In London, where Gall Zeidler Consultants has been working on Crossrail, the tunnels being dug beneath the capital will eventually increase the city's total public transport capacity by 10 per cent, which will significantly reduce congestion and reliance on cars and improve the commuters' experience.

Additionally, engineers can go underground to improve existing infrastructure and Gall Zeidler has helped increase the capacity of several underground stations in London, including King's Cross, Tottenham Court Road and Bond Street. Zeidler says that tunnels can even be used to house potentially hazardous pieces of infrastructure, such as treatment plants, as is happening in congested cities like Singapore and Hong Kong.

It's essential, however, that these tunnels are built properly, using the cleanest and best technology and materials available. That way, tunnels will last longer and work better. Proper waterproofing should be used wherever possible to eliminate serious issues such as groundwater seepage, which can cause structural and operational damage as it has in several underground schemes around the world, where Gall Zeidler is rehabilitating older tunnels. Such work is necessary but requires considerable expenditure, energy and effort, all of which cause disruption. Tunnellers therefore need to think about the impact of their work on the world around them, including the amount of traffic the work will generate, and they especially need to plan for future interventions, facilitating renovation and restoration to take place to give the structures an extended life.

"As tunnellers, we have many opportunities to use our technologies to reduce the impact on our environment," says Zeidler. "However, I do not believe that sustainability stops at avoiding the use of harmful technologies and materials; it should embrace the durability of our products. Durability includes longevity but also facilitating economic and ecological maintenance and repair to extend the lifetime of our structures beyond the typical design life. You need to ensure that the structures that are being built today are more durable."

Measures can be taken during design and construction that will increase the life of a tunnel and make ongoing maintenance easier and cheaper. However, this will usually cost more, meaning clients have to be prepared to make the required investment upfront – and that can only happen if engineers understand and communicate the costs and benefits to clients. "As engineers, we need to point out the long-term gains when it comes to maintenance and operation," says Zeidler. "It's a small contribution but if we all take that seriously we can make a difference, and tunnels represent a great opportunity for sustainability."

APPENDICES

Sustainable Development Goal 1

End poverty in all forms everywhere

Goal 1 commits to ending poverty in all its manifestations, including its most extreme forms, over the next 10 years, and resolves that all people, everywhere, should enjoy a basic standard of living. This includes social protection benefits for the poor and most vulnerable and ensuring that people harmed by conflict and natural hazards receive adequate support, including access to basic services.

The multi-dimensional definition of poverty extends beyond a measure of financial income and is characterized by severe deprivation of all basic human needs. Infrastructure can address poverty reduction by providing of a range of basic services at the household level, including fuel for heating, lighting and cooking, water for drinking and sanitation, and adequate disposal of waste. Building networks of infrastructure will facilitate the provision of these services directly to households and communities in need. In rural or remote areas, improved transportation networks and digital communications can provide access to these services, supplying residents with vital goods and information and linking them to facilities such as schools or hospitals.

Access is key to improving resilience in the face of extreme climatic events, or in the case of war and conflict. The sustained provision of services such as electricity, water and waste disposal will ensure that communities have the means to recover swiftly from social or environmental shocks without falling back into poverty. The establishment of social protection systems for the poor can also improve resilience. This requires information management systems that are most effectively implemented through digital means.

TARGETS

1.1 By 2030, eradicate extreme poverty for all people everywhere, currently measured as people living on less than $1.25 a day

1.2 By 2030, reduce at least by half the proportion of men, women and children of all ages living in poverty in all its dimensions according to national definitions

1.3 Implement nationally appropriate social protection systems and measures for all, including floors, and by 2030 achieve substantial coverage of the poor and the vulnerable

1.4 By 2030, ensure that all men and women, in particular the poor and the vulnerable, have equal rights to economic resources, as well as access to basic services, ownership and control over land and other forms of property, inheritance, natural resources, appropriate new technology and financial services, including microfinance

1.5 By 2030, build the resilience of the poor and those in vulnerable situations and reduce their exposure and vulnerability to climate-related extreme events and other economic, social and environmental shocks and disasters

1.A Ensure significant mobilization of resources from a variety of sources, including through enhanced development cooperation, in order to provide adequate and predictable means for developing countries, in particular least developed countries, to implement programmes and policies to end poverty in all its dimensions

1.B Create sound policy frameworks at the national, regional and international levels, based on pro-poor and gender-sensitive development strategies, to support accelerated investment in poverty eradication actions

Sustainable Development Goal 2

End hunger, achieve food security and improved nutrition and promote sustainable agriculture

Goal 2 seeks sustainable solutions to end hunger in all its forms by 2030 and to achieve food security, aiming to ensure that everyone everywhere has enough good-quality food to lead a healthy life. Achieving this goal will require better access to food and the widespread promotion of sustainable agriculture.

Ending hunger through sustainable food systems should account for each step in the chain of food production and consumption within a systems context: growing, processing, distribution, preparation, consumption and disposal of wastes. Through infrastructure contributions in all sectors, the secure provision of food and nutrition can be assured for all segments of society.

Improved water supply and irrigation techniques are key to enhancing agricultural resilience to drought and maintaining food supply, while agricultural land use management can improve resilience to flooding. Modern energy infrastructure plays a key role in providing efficiency to the production process: electrification of food processing can increase productivity through automation, while refrigeration can contribute toward the reduction of food waste and enhance food quality and freshness. The construction of warehouses and other storage facilities help to manage the food production process and reduce waste throughout the supply chain.

More efficient transportation that links producers and consumers will increase the sustainability of the food system and increase the variety of food options available to consumers. Digital communications can play a larger role in food distribution, as producers gain access to market information both domestically and internationally, and provides farmers with the knowledge necessary to improve agricultural techniques.

TARGETS

2.1 By 2030, end hunger and ensure access by all people, in particular the poor and people in vulnerable situations, including infants, to safe, nutritious and sufficient food all year round

2.2 By 2030, end all forms of malnutrition, including achieving, by 2025, the internationally agreed targets on stunting and wasting in children under 5 years of age, and address the nutritional needs of adolescent girls, pregnant and lactating women and older persons

2.3 By 2030, double the agricultural productivity and incomes of small-scale food producers, in particular women, indigenous peoples, family farmers, pastoralists and fishers, including through secure and equal access to land, other productive resources and inputs, knowledge, financial services, markets and opportunities for value addition and non-farm employment

2.4 By 2030, ensure sustainable food production systems and implement resilient agricultural practices that increase productivity and production, that help maintain ecosystems, that strengthen capacity for adaptation to climate change, extreme weather, drought, flooding and other disasters and that progressively improve land and soil quality

2.5 By 2020, maintain the genetic diversity of seeds, cultivated plants and farmed and domesticated animals and their related wild species, including through soundly managed and diversified seed and plant banks at the national, regional and international levels, and promote access to and fair and equitable sharing of benefits arising from the utilization of genetic resources and associated traditional knowledge, as internationally agreed

2.A Increase investment, including through enhanced international cooperation, in rural infrastructure, agricultural research and extension services, technology development and plant and livestock gene banks in order to enhance agricultural productive capacity in developing countries, in particular least developed countries

2.B Correct and prevent trade restrictions and distortions in world agricultural markets, including through the parallel elimination of all forms of agricultural export subsidies and all export measures with equivalent effect, in accordance with the mandate of the Doha Development Round

2.A Adopt measures to ensure the proper functioning of food commodity markets and their derivatives and facilitate timely access to market information, including on food reserves, in order to help limit extreme food price volatility

Sustainable Development Goal 3

Ensure healthy lives and promote well-being for all at all ages

Goal 3 seeks to ensure health and well-being for all by improving reproductive, maternal and child health, ending disease epidemics, and ensuring health coverage and access to safe, affordable and effective medicines and vaccines. Towards that end, world leaders committed to support research and development, increase health financing, and strengthen the capacity of all countries to reduce and manage health risks.

Most of the world's population, particularly in rapidly growing urban areas, will increasingly benefit from access to health services provided in hospitals and clinics, while universities and other educational establishments support medical research, development and training. These facilities require reliable networks of energy, water and digital communications infrastructure, which, together, provide the necessary conditions for the functioning of advanced medical equipment and technology. Sanitation services through clean water provision and waste disposal can limit the spread of diseases, including neglected tropical diseases scheduled for elimination by the WHO, and have been found to greatly decrease the incidence of maternal, neonatal and child mortality by reducing the spread of water-borne pathogens.

Where access to these facilities is not possible, such as in rural or remote communities, transportation networks are crucial to the provision of health services, including prevention through education and public awareness. In an increasingly connected world, information and communication technologies can provide access to health services by facilitating the sharing of knowledge, records and results and assisting in self-management of medical conditions, including addiction. Digital communications infrastructure can provide an array of additional health-related benefits to society, including the recruitment and training of the health workforce and the timely dissemination of information to reduce health risks from natural disasters.

TARGETS

3.1 By 2030, reduce the global maternal mortality ratio to less than 70 per 100,000 live births

3.2 By 2030, end preventable deaths of newborns and children under 5 years of age, with all countries aiming to reduce neonatal mortality to at least as low as 12 per 1,000 live births and under-5 mortality to at least as low as 25 per 1,000 live births

3.3 By 2030, end the epidemics of AIDS, tuberculosis, malaria and neglected tropical diseases and combat hepatitis, water-borne diseases and other communicable diseases

3.4 By 2030, reduce by one third premature mortality from non-communicable diseases through prevention and treatment and promote mental health and well-being

3.5 Strengthen the prevention and treatment of substance abuse, including narcotic drug abuse and harmful use of alcohol

3.6 By 2020, halve the number of global deaths and injuries from road traffic accidents

3.7 By 2030, ensure universal access to sexual and reproductive health-care services, including for family planning, information and education, and the integration of reproductive health into national strategies and programmes

3.8 Achieve universal health coverage, including financial risk protection, access to quality essential health-care services and access to safe, effective, quality and affordable essential medicines and vaccines for all

3.9 By 2030, substantially reduce the number of deaths and illnesses from hazardous chemicals and air, water and soil pollution and contamination

3.A Strengthen the implementation of the World Health Organization Framework Convention on Tobacco Control in all countries, as appropriate

3.B Support the research and development of vaccines and medicines for the communicable and non-communicable diseases that primarily affect developing countries, provide access to affordable essential medicines and vaccines, in accordance with the Doha Declaration on the TRIPS Agreement and Public Health, which affirms the right of developing countries to use to the full the provisions in the Agreement on Trade-Related Aspects of Intellectual Property Rights regarding flexibilities to protect public health, and, in particular, provide access to medicines for all

3.C Substantially increase health financing and the recruitment, development, training and retention of the health workforce in developing countries, especially in least developed countries and small island developing States

3.D Strengthen the capacity of all countries, in particular developing countries, for early warning, risk reduction and management of national and global health risks

Sustainable Development Goal 4

Ensure inclusive and equitable quality education and promote lifetime opportunities for all

Goal 4 aims to ensure that all people have access to quality education and lifelong learning opportunities, focusing on the acquisition of foundational and higher-order skills at all stages of education and development; greater and more equitable access to quality education at all levels as well as technical and vocational education and training; and the knowledge, skills and values needed to function well and contribute to society. In addition to classrooms, desks and learning materials, educational facilities such as schools, colleges and universities require a range of infrastructure services to provide quality learning opportunities and skills development. The performance of students and pupils can be greatly enhanced through the provision of electricity for heating and lighting, water and sanitation facilities, and, importantly, access to digital infrastructure such as computers and the internet.

As well as increasing the knowledge capacity of society as a whole, internet access provides teachers with a range of educational materials that can be accessed online or downloaded, and likewise used to improve instructors' skills and teaching methods. Access to ICT infrastructure at all levels of schooling fosters digital skill development, which is increasingly important for employment and entrepreneurship opportunities. Youth and adults who have relevant skills face better prospects in job markets that increasingly value technical or vocational proficiency. Digital communications can also provide access to learning for young people in villages or rural communities who are unable to travel far from home. The availability of suitable transportation can fulfil a similar purpose, enabling students to progress to higher education in another city or region. Many young children, especially girls, are assigned domestic responsibilities such as the collection of drinking water or firewood, leaving them no time to attend school. Modern and accessible energy and water infrastructure can thus provide time savings for families and help ensure all children have the chance to pursue their education.

TARGETS

4.1 By 2030, ensure that all girls and boys complete free, equitable and quality primary and secondary education leading to relevant and Goal-4 effective learning outcomes

4.2 By 2030, ensure that all girls and boys have access to quality early childhood development, care and preprimary education so that they are ready for primary education

4.3 By 2030, ensure equal access for all women and men to affordable and quality technical, vocational and tertiary education, including university

4.4 By 2030, substantially increase the number of youth and adults who have relevant skills, including technical and vocational skills, for employment, decent jobs and entrepreneurship

4.5 By 2030, eliminate gender disparities in education and ensure equal access to all levels of education and vocational training for the vulnerable, including persons with disabilities, indigenous peoples and children in vulnerable situations

4.6 By 2030, ensure that all youth and a substantial proportion of adults, both men and women, achieve literacy and numeracy

4.7 By 2030, ensure that all learners acquire the knowledge and skills needed to promote sustainable development, including, among others, through education for sustainable development and sustainable lifestyles, human rights, gender equality, promotion of a culture of peace and non-violence, global citizenship and appreciation of cultural diversity and of culture's contribution to sustainable development

4.A Build and upgrade education facilities that are child, disability and gender sensitive and provide safe, nonviolent, inclusive and effective learning environments for all

4.B By 2020, substantially expand globally the number of scholarships available to developing countries, in particular least developed countries, small island developing States and African countries, for enrolment in higher education, including vocational training and information and communications technology, technical, engineering and scientific programmes, in developed countries and other developing countries

4.C By 2030, substantially increase the supply of qualified teachers, including through international cooperation for teacher training in developing countries, especially least developed countries and small island developing states

Sustainable Development Goal 5

Achieve gender equality and empower all women and girls

Empowering women and girls to reach their full potential requires that they have equal opportunities to those of men and boys. This means eliminating all forms of discrimination and violence against them, including violence by intimate partners, sexual violence and harmful practices, such as child marriage and female genital mutilation (FGM). Ensuring that women have better access to paid employment, sexual and reproductive health and reproductive rights, and real decision-making power in public and private spheres will further ensure that development is equitable and sustainable.

Infrastructure's role in achieving targets on gender equality is largely linked to creating opportunities for women's empowerment by facilitating and modernising the provision of infrastructure services that have been traditionally assigned to women. In many regions, particularly in rural areas, girls and women may spend hours each day collecting fuel, such as firewood, for cooking and household use, as well as drinking water. As a result, many are limited in their educational or economic potential. The provision of accessible energy and water supply infrastructure in all communities can allow more time for the equitable pursuit of economic, social, and leadership activities, and reduce time spent in unpaid domestic work. Information and communication infrastructure provides enabling technology by which women may attain economic empowerment or greater influence in their communities.

In addition to time savings, women's health, safety and well-being can be improved through the provision of infrastructure. Electrification may reduce exposure to cooking fumes in homes, while street lighting may decrease the potential for sexual harassment and violence towards women and girls in public spaces. Transportation allows the provision and administration of sexual and reproductive health supplies and services to all communities, including inaccessible or remote areas. More broadly, eliminating discriminatory practices and gender-based violence requires infrastructure related to governance and rule of law at the national and local level to implement legal protections, enforce laws and ensure access to justice.

TARGETS

5.1 End all forms of discrimination against all women and girls everywhere

5.2 Eliminate all forms of violence against all women and girls in the public and private spheres, including trafficking and sexual and other types of exploitation

5.3 Eliminate all harmful practices, such as child, early and forced marriage and female genital mutilation

5.4 Recognize and value unpaid care and domestic work through the provision of public services, infrastructure and social protection policies and the promotion of shared responsibility within the household and the family as nationally appropriate

5.5 Ensure women's full and effective participation and equal opportunities for leadership at all levels of decisionmaking in political, economic and public life

5.6 Ensure universal access to sexual and reproductive health and reproductive rights as agreed in accordance with the Programme of Action of the International Conference on Population and Development and the Beijing Platform for Action and the outcome documents of their review conferences

5.A Undertake reforms to give women equal rights to economic resources, as well as access to ownership and control over land and other forms of property, financial services, inheritance and natural resources, in accordance with national laws

5.B Enhance the use of enabling technology, in particular information and communications technology, to promote the empowerment of women

5.C Adopt and strengthen sound policies and enforceable legislation for the promotion of gender equality and the empowerment of all women and girls at all levels

Sustainable Development Goal 6

The UN's goal to "ensure availability and sustainable management of water and sanitation for all" in depth

Goal 6 goes beyond drinking water, sanitation and hygiene to also address the quality and sustainability of water resources, critical to the survival of people and the planet. The 2030 agenda recognises the centrality of water resources to sustainable development, and the vital role that improved drinking water, sanitation and hygiene play in progress in other areas, including health, education and poverty reduction.

An integrated water resources management approach to water security promotes co-ordinated development and management of water and related resources to maximise social, economic and environmental outcomes. Within this context, investment in water supply, wastewater and flood risk management infrastructure is required for the achievement of all Goal 6 outcome targets (6.1 to 6.6). This includes universal and equitable access to safe and affordable drinking water and sanitation services, managed abstraction and discharge and the elimination of solid and hazardous waste dumping to protect water-related ecosystems. Better water use efficiency through reduced leakage and increased reuse can maintain freshwater withdrawals at sustainable levels.

Energy systems often form a component of water supply, such as through the operation of pumps. They can also be water-intensive and should therefore be designed to minimize water usage and pollution. In certain regions, the delivery of water by means of trucks may be the more feasible option; thus, a reliable road network to access these communities can provide an alternative to the construction of piped water networks, which may be financially burdensome.

Given the transboundary nature of many rivers, lakes, aquifers and waterways, international cooperation and capacity building are important components of effective water management. This requires governance and institutional infrastructure, including community-level venues, where stakeholders can access information and receive training and support from local or international experts and practitioners. Transportation networks and the use of digital communications infrastructure provide the access and capabilities to further these objectives.

TARGETS

6.1 By 2030, achieve universal and equitable access to safe and affordable drinking water for all

6.2 By 2030, achieve access to adequate and equitable sanitation and hygiene for all and end open defecation, paying special attention to the needs of women and girls and those in vulnerable situations

6.3 By 2030, improve water quality by reducing pollution, eliminating dumping and minimizing release of hazardous chemicals and materials, halving the proportion of untreated wastewater and substantially increasing recycling and safe reuse globally

6.4 By 2030, substantially increase water-use efficiency across all sectors and ensure sustainable withdrawals and supply of freshwater to address water scarcity and substantially reduce the number of people suffering from water scarcity

6.5 By 2030, implement integrated water resources management at all levels, including through transboundary cooperation as appropriate

6.6 By 2020, protect and restore water-related ecosystems, including mountains, forests, wetlands, rivers, aquifers and lakes

6.A By 2030, expand international cooperation and capacity-building support to developing countries in water- and sanitation-related activities and programmes, including water harvesting, desalination, water efficiency, wastewater treatment, recycling and reuse technologies

6.B Support and strengthen the participation of local communities in improving water and sanitation management

Sustainable Development Goal 7

Ensure access to affordable, reliable, sustainable and modern energy for all

Access to affordable, reliable and sustainable energy is crucial to achieving many of the Sustainable Development Goals. Energy access, however, varies widely across countries and the current rate of progress falls short of what will be required to achieve the Goal. Redoubled efforts will be needed, particularly for countries with large energy access deficits and high energy consumption.

Given its vital role as a resource input to development, infrastructure will need to address core dimensions of energy sustainability, defined by the World Energy Council as the "Trilemma" of energy security, universal access to affordable energy services, and environmentally sensitive production and use of energy. The outcome targets of Goal 7 (7.1 -7.3) capture each of these broader objectives and highlight the important interdependence between the energy and transport sectors, where improved fuel efficiency in vehicles and technological trends such as the electrification of transport may simultaneously reduce energy demand and fossil fuel emissions on a large scale if made more widely accessible to the public

As with other forms of infrastructure growth, technological innovation in the energy sector with large-scale global impacts will require global cooperation, knowledge transfer and capacity building through research and investment flows, which may be enhanced through the involvement of higher education and research facilities. Digital technology and communications will also be a key input to growth in this sector.

TARGETS

7.1 By 2030, ensure universal access to affordable, reliable and modern energy services

7.2 By 2030, increase substantially the share of renewable energy in the global energy mix

7.3 By 2030, double the global rate of improvement in energy efficiency

7.A By 2030, enhance international cooperation to facilitate access to clean energy research and technology, including renewable energy, energy efficiency and advanced and cleaner fossil-fuel technology, and promote investment in energy infrastructure and clean energy technology

7.B By 2030, expand infrastructure and upgrade technology for supplying modern and sustainable energy services for all in developing countries, in particular least developed countries, small island developing States, and land-locked developing countries, in accordance with their respective programmes of support

Sustainable Development Goal 8

Promote sustained, inclusive and sustainable economic growth, full and productive employment and decent work for all

Sustained and inclusive economic growth is a prerequisite for sustainable development, which can contribute to improved livelihoods for people around the world. Economic growth can lead to new and better employment opportunities and provide greater economic security for all. Moreover, rapid growth, especially among the least developed and other developing countries, can help them reduce the wage gap relative to developed countries, thereby diminishing glaring inequalities between the rich and poor. Infrastructure systems play an important role in increasing national economic growth and productivity, which require secure and accessible energy sources, transport networks to link producers and consumers, digital technology and communications to increase value added and provide efficiencies at all stages of the value chain. Water is a key input to manufacturing, while waste management systems treat or recover energy from billions of tonnes of residual materials created from the production process. Access to reliable infrastructure that performs these functions, as well as the provision of built economic infrastructure such as factories, industrial and storage facilities and markets, is key to attracting new investment and expanding existing economic activity while generating employment and improving livelihoods. Moreover, adapting infrastructure to function at decreased levels of resource use and incorporating user behaviour in product design can ensure that this growth is sustained without depleting the world's natural resources or inflicting irreversible damage on the environment.

Making economic growth more inclusive means reaching and providing economic opportunity to all, including the most vulnerable. Improved transportation networks can link workers with suitable jobs, facilitating rural to urban and interurban commuting for skilled and unskilled workers, including women and youth. The transition to a digital economy creates new opportunities for growth through a range of ICT-based jobs while allowing for teleworking and financial inclusion through access to digital banking and services.

TARGETS

8.1 Sustain per capita economic growth in accordance with national circumstances and, in particular, at least 7 per cent gross domestic product growth per annum in the least developed countries

8.2 Achieve higher levels of economic productivity through diversification, technological upgrading and innovation, including through a focus on high-value added and labour-intensive sectors

8.3 Promote development-oriented policies that support productive activities, decent job creation, entrepreneurship, creativity and innovation, and encourage the formalization and growth of micro-, small- and medium-sized enterprises, including through access to financial services

8.4 Improve progressively, through 2030, global resource efficiency in consumption and production and endeavour to decouple economic growth from environmental degradation, in accordance with the 10-year framework of programmes on sustainable consumption and production, with developed countries taking the lead

8.5 By 2030, achieve full and productive employment and decent work for all women and men, including for young people and persons with disabilities, and equal pay for work of equal value

8.6 By 2020, substantially reduce the proportion of youth not in employment, education or training

8.7 Take immediate and effective measures to eradicate forced labour, end modern slavery and human trafficking and secure the prohibition and elimination of the worst forms of child labour, including recruitment and use of child soldiers, and by 2025 end child labour in all its forms

8.8 Protect labour rights and promote safe and secure working environments for all workers, including migrant workers, in particular women migrants, and those in precarious employment

8.9 By 2030, devise and implement policies to promote sustainable tourism that creates jobs and promotes local culture and products

8.10 Strengthen the capacity of domestic financial institutions to encourage and expand access to banking, insurance and financial services for all

8.A Increase Aid for Trade support for developing countries, in particular least developed countries, including through the Enhanced Integrated Framework for Trade-Related Technical Assistance to Least Developed Countries

8.B By 2020, develop and operationalize a global strategy for youth employment and implement the Global Jobs Pact of the International Labour Organization

Sustainable Development Goal 9

Build resilient infrastructure, promote inclusive and sustainable industrialization and foster innovation

Goal 9 addresses three important aspects of sustainable development: infrastructure, industrialization and innovation. Infrastructure provides the basic physical facilities essential to business and society; industrialization drives economic growth and job creation, thereby reducing income inequality; and innovation expands the technological capabilities of industrial sectors and leads to the development of new skills. By definition, achieving Goal 9 targets requires growth across all sectors to strengthen the resilience, inclusiveness and sustainability of infrastructure systems. The benefits of strong national infrastructure extend to the achievement of outcomes across all 17 Sustainable Development Goals. For example, the design of infrastructure systems can specifically incorporate objectives around industrialisation and industry (Goal 8), sustainability and resource-use efficiency (Goals 7 and 12), and equitable access (Goals 5 and 10). Future infrastructure development will need to address questions around the balance of capacity provision and demand management, vulnerabilities in infrastructure networks, and governance models used to finance and deliver infrastructure services.

The technological innovation required to progress toward Goal 9 will involve extensive R&D and scientific research, in addition to financial and technical support to developing countries. Access to information and communication technology will play a large role in fostering the international cooperation required and skills development required to achieve this level of innovation.

TARGETS

9.1 Develop quality, reliable, sustainable and resilient infrastructure, including regional and transborder infrastructure, to support economic development and human well-being, with a focus on affordable and equitable access for all

9.2 Promote inclusive and sustainable industrialization and, by 2030, significantly raise industry's share of employment and gross domestic product, in line with national circumstances, and double its share in least developed countries

9.3 Increase the access of small-scale industrial and other enterprises, in particular in developing countries, to financial services, including affordable credit, and their integration into value chains and markets

9.4 By 2030, upgrade infrastructure and retrofit industries to make them sustainable, with increased resource-use efficiency and greater adoption of clean and environmentally sound

technologies and industrial processes, with all countries taking action in accordance with their respective capabilities

9.5 Enhance scientific research, upgrade the technological capabilities of industrial sectors in all countries, in particular developing countries, including, by 2030, encouraging innovation and substantially increasing the number of research and development workers per 1 million people and public and private research and development spending

9.A Facilitate sustainable and resilient infrastructure development in developing countries through enhanced financial, technological and technical support to African countries, least developed countries, landlocked developing countries and small island developing States

9.B Support domestic technology development, research and innovation in developing countries, including by ensuring a conducive policy environment for, inter alia, industrial diversification and value addition to commodities

9.C Significantly increase access to information and communications technology and strive to provide universal and affordable access to the Internet in least developed countries by 2020

Sustainable Development Goal 10

Reduce inequality within and among countries

Goal 10 calls for reducing inequalities in income as well as those based on sex, age, disability, race, class, ethnicity, religion and opportunity – both within and among countries – and addresses issues related to representation and development assistance. The provision of infrastructure has been shown to reduce inequality within countries. Among the poorest, increased access to infrastructure can improve health and well-being by providing reliable basic services and allowing people to pursue livelihoods and economic opportunities. Built infrastructure, including shelter and community governance facilities as well as energy, water and waste treatment networks, should be designed to target particularly vulnerable segments of society.

An example of infrastructure's potential role in empowering disadvantaged groups is highlighted in Goal 5: due to traditional gender roles, domestic responsibilities including the collection of fuel and water may limit the economic and educational potential of women and girls in some regions. This disadvantage can be removed through the provision of modern energy or water supply infrastructure, or, alternatively, efficient transportation allowing for the delivery of these services directly to communities and households.

Infrastructure can also reduce inequalities between countries by encouraging financial flows, through foreign investment, to regions where needs are greatest. A country's attractiveness to investment is increased with quality and resilient infrastructure systems by lowering costs of doing business and smoothing the functioning of investors' production and trade activities. Digital payment methods provide more efficient channels for transferring migrant remittances to family members in countries of origin.

TARGETS

10.1 By 2030, progressively achieve and sustain income growth of the bottom 40 per cent of the population at a rate higher than the national average

10.2 By 2030, empower and promote the social, economic and political inclusion of all, irrespective of age, sex, disability, race, ethnicity, origin, religion or economic or other status

10.3 Ensure equal opportunity and reduce inequalities of outcome, including by eliminating discriminatory laws, policies and practices and promoting appropriate legislation, policies and action in this regard

10.4 Adopt policies, especially fiscal, wage and social protection policies, and progressively achieve greater equality

10.5 Improve the regulation and monitoring of global financial markets and institutions and strengthen the implementation of such regulations

10.6 Ensure enhanced representation and voice for developing countries in decision-making in global international economic and financial institutions in order to deliver more effective, credible, accountable and legitimate institutions

10.7 Facilitate orderly, safe, regular and responsible migration and mobility of people, including through the implementation of planned and well-managed migration policies

10.A Implement the principle of special and differential treatment for developing countries, in particular least developed countries, in accordance with World Trade Organization agreements

10.B Encourage official development assistance and financial flows, including foreign direct investment, to States where the need is greatest, in particular least developed countries, African countries, small island developing States and landlocked developing countries, in accordance with their national plans and programmes

10.C By 2030, reduce to less than 3 per cent the transaction costs of migrant remittances and eliminate remittance corridors with costs higher than 5 per cent

Sustainable Development Goal 11

Make cities and human settlements inclusive, safe, resilient and sustainable

Today, more than half the world's population lives in cities. By 2030, it is projected that 6 in 10 people will be urban dwellers. Despite numerous planning challenges, cities offer more efficient economies of scale on many levels, including the provision of goods, services and transportation. With sound, risk-informed planning and management, cities can become incubators for innovation and growth and drivers of sustainable development.

With urban areas estimated to grow by two-and-a-half times by 2050, extensive cross-sectoral infrastructure planning for cities will need to be implemented to ensure basic services for urban dwellers, including health, educational, housing and other facilities that require energy, water, transportation and digital communication networks as well as effective waste management infrastructure.

New urban infrastructure must be built to be safe and affordable for its residents. Transport systems can be designed, using increased digital technology, to reduce road accidents, while public transportation and sanitation facilities can be expanded to provide safer and accessible options to all people. Air quality, linked to emissions from the energy and transport sectors, as well as municipal waste and wastewater management require specific attention in cities to reduce or eliminate adverse health effects.

In the context of climate change, and given the economic and social costs at stake, resiliency must be built into new and existing critical infrastructure in cities across all sectors as outlined in the Sendai Framework for disaster risk reduction. Adequate flood risk infrastructure will ensure additional protection for cities' cultural heritage.

TARGETS

11.1 By 2030, ensure access for all to adequate, safe and affordable housing and basic services and upgrade slums

11.2 By 2030, provide access to safe, affordable, accessible and sustainable transport systems for all, improving road safety, notably by expanding public transport, with special attention to the needs of those in vulnerable situations, women, children, persons with disabilities and older persons

11.3 By 2030, enhance inclusive and sustainable urbanization and capacity for participatory, integrated and sustainable human settlement planning and management in all countries

11.4 Strengthen efforts to protect and safeguard the world's cultural and natural heritage

11.5 By 2030, significantly reduce the number of deaths and the number of people affected and substantially decrease the direct economic losses relative to global gross domestic product caused by disasters, including water-related disasters, with a focus on protecting the poor and people in vulnerable situations

11.6 By 2030, reduce the adverse per capita environmental impact of cities, including by paying special attention to air quality and municipal and other waste management

11.7 By 2030, provide universal access to safe, inclusive and accessible, green and public spaces, in particular for women and children, older persons and persons with disabilities

11.A Support positive economic, social and environmental links between urban, peri-urban and rural areas by strengthening national and regional development planning

11.B By 2020, substantially increase the number of cities and human settlements adopting and implementing integrated policies and plans towards inclusion, resource efficiency, mitigation and adaptation to climate change, resilience to disasters, and develop and implement, in line with the Sendai Framework for Disaster Risk Reduction 2015-2030, holistic disaster risk management at all levels

11.C Support least developed countries, including through financial and technical assistance, in building sustainable and resilient buildings utilizing local materials

Sustainable Development Goal 12

Ensure sustainable consumption and production patterns

Sustainable growth and development require minimizing the natural resources and toxic materials used, and the waste and pollutants generated, throughout the entire production and consumption process. Goal 12 encourages more sustainable consumption and production patterns through various measures, including specific policies and international agreements on the management of materials that are toxic to the environment.

Consumption and production rely on a continued supply of resource inputs, including energy and water, as well as the treatment of waste outputs to the air, water and soil. Redesigning these processes to reduce resource use and minimise environmental impacts will involve fundamental changes to the way in which we plan, construct and use infrastructure systems.

Food waste occurs at all stages of the supply chain and accounts for one-third of total food produced for human consumption, or approximately 1.3 billion tonnes per year. The provision of adequate refrigeration throughout the production process, improvements to the efficiency of food transportation and the use of digital technology to improve production and consumption choices can assist in limiting these losses.

The use of digital communications can influence behaviour to reduce the environmental impacts of consumption. With computers and mobile phones in widespread use, the internet can provide useful resources to promote behavioural changes such as reduced energy consumption, recycling and general sustainability awareness. These can also be incorporated in school curricula or sustainable tourism campaigns. The increasing substitution of in-person meetings with online, phone or video communication may contribute to decreased traffic congestion and transport emissions.

TARGETS

12.1 Implement the 10-year framework of programmes on sustainable consumption and production, all countries taking action, with developed countries taking the lead, taking into account the development and capabilities of developing countries

12.2 By 2030, achieve the sustainable management and efficient use of natural resources

12.3 By 2030, halve per capita global food waste at the retail and consumer levels and reduce food losses along production and supply chains, including post-harvest losses

12.4 By 2020, achieve the environmentally sound management of chemicals and all wastes throughout their life cycle, in accordance with agreed international frameworks, and significantly reduce their release to air, water and soil in order to minimize their adverse impacts on human health and the environment

12.5 By 2030, substantially reduce waste generation through prevention, reduction, recycling and reuse

12.6 Encourage companies, especially large and transnational companies, to adopt sustainable practices and to integrate sustainability information into their reporting cycle

12.7 Promote public procurement practices that are sustainable, in accordance with national policies and priorities

12.8 By 2030, ensure that people everywhere have the relevant information and awareness for sustainable development and lifestyles in harmony with nature

12.A Support developing countries to strengthen their scientific and technological capacity to move towards more sustainable patterns of consumption and production

12.B Develop and implement tools to monitor sustainable development impacts for sustainable tourism that creates jobs and promotes local culture and products

12.C Rationalize inefficient fossil-fuel subsidies that encourage wasteful consumption by removing market distortions, in accordance with national circumstances, including by restructuring taxation and phasing out those harmful subsidies, where they exist, to reflect their environmental impacts, taking fully into account the specific needs and conditions of developing countries and minimizing the possible adverse impacts on their development in a manner that protects the poor and the affected communities

Sustainable Development Goal 13

Take urgent action to combat climate change and its impacts

Climate change presents the single biggest threat to development, and its widespread, unprecedented effects disproportionately burden the poorest and the most vulnerable. Goal 13 calls for urgent action not only to combat climate change and its impacts, but also to build resilience in responding to climate-related hazards and natural disasters. Efforts to implement climate change adaptation and disaster risk reduction, as formalised in the 2015 Sendai Framework, highlight the need to invest in and enhance disaster preparedness and resilience to climate-related hazards. To this end, infrastructure planning across all sectors should be integrated in national policies to ensure that the most vulnerable communities do not suffer disproportionate impacts from climate change.

In coastal regions and small island states, increasing sea level rise and greater occurrence of storm surges necessitate extensive flood risk management infrastructure to reduce exposure to water-related hazards. The design of energy, water supply and waste management systems should be such that affected communities can continue to receive basic services in the event of a disaster, including fuel and adequate sanitation, which will reduce the incidence of epidemics and other social risks. Should a community suffer extensive damage to key infrastructure assets, including homes, transportation links will be crucial for providing emergency aid during the recovery process.

In addition to providing early warning capabilities, digital information systems contribute to countries' capacities to better understand disaster risk and to strengthen responses through data collection, research and public awareness-raising.

TARGETS

13.1 Strengthen resilience and adaptive capacity to climate-related hazards and natural disasters in all countries

13.2 Integrate climate change measures into national policies, strategies and planning

13.3 Improve education, awareness-raising and human and institutional capacity on climate change mitigation, adaptation, impact reduction and early warning

13.A Implement the commitment undertaken by developed-country parties to the United Nations Framework Convention on Climate Change to a goal of mobilizing jointly $100 billion annually by 2020 from all sources to address the needs of developing countries in the context of meaningful mitigation actions and transparency on implementation and fully operationalize the Green Climate Fund through its capitalization as soon as possible

13.B Promote mechanisms for raising capacity for effective climate change-related planning and management in least developed countries and small island developing States, including focusing on women, youth and local and marginalized communities

Acknowledging that the United Nations Framework Convention on Climate Change is the primary international, intergovernmental forum for negotiating the global response to climate change.

Sustainable Development Goal 14

Conserve and sustainably use the oceans, seas and marine resources for sustainable development

Oceans, seas and other marine resources are essential to human well-being and social and economic development worldwide. They provide livelihoods, subsistence and benefits from fisheries, tourism and other sectors, while regulating the global ecosystem by absorbing heat and carbon dioxide from the atmosphere. However, oceans and coastal areas are extremely vulnerable to environmental degradation, overfishing, climate change and pollution.

Eliminating the pollution of oceans, seas and marine areas will require action across infrastructure sectors: connecting cities and communities to wastewater treatment plants to prevent untreated discharge in waterways; providing adequate solid waste and recycling facilities to eliminate the emission of plastics and other wastes; incorporating coastal erosion in flood risk management plans; and regulating marine transport and contamination from fuels which are a major source of marine pollution.

On a biodiversity level, a global transition away from fossil fuel burning and toward low-carbon energy infrastructure is necessary to combat ocean acidification, which, by reducing the availability of carbonate minerals in seawater, disrupts large components of the marine food web. Preventing further biodiversity loss due to overfishing may also increasingly incorporate digital technologies such as satellite tracking and remote monitoring to ensure traceability of fish harvests and enforcement of penalties in line with international treaties. Such actions are required to preserve the livelihoods of small-scale fishers, who also benefit from energy, transport and digital infrastructure to market their products.

Conservation of the oceans and their resources requires effective and participatory governance through international cooperation. Digital communications and technology can assist in providing the knowledge transfer and information sharing necessary to link the implementation of actions and policies between international partners.

TARGETS

14.1 By 2025, prevent and significantly reduce marine pollution of all kinds, in particular from land-based activities, including marine debris and nutrient pollution

14.2 By 2020, sustainably manage and protect marine and coastal ecosystems to avoid significant adverse impacts, including by strengthening their resilience, and take action for their restoration in order to achieve healthy and productive oceans

14.3 Minimize and address the impacts of ocean acidification, including through enhanced scientific cooperation at all levels

14.4 By 2020, effectively regulate harvesting and end overfishing, illegal, unreported and unregulated fishing and destructive fishing practices and implement science-based management plans, in order to restore fish stocks in the shortest time feasible, at least to levels that can produce maximum sustainable yield as determined by their biological characteristics

14.5 By 2020, conserve at least 10 per cent of coastal and marine areas, consistent with national and international law and based on the best available scientific information

14.6 By 2020, prohibit certain forms of fisheries subsidies which contribute to overcapacity and overfishing, eliminate subsidies that contribute to illegal, unreported and unregulated fishing and refrain from introducing new such subsidies, recognizing that appropriate and effective special and differential treatment for developing and least developed countries should be an integral part of the World Trade Organization fisheries subsidies negotiation

14.7 By 2030, increase the economic benefits to Small Island developing States and least developed countries from the sustainable use of marine resources, including through sustainable management of fisheries, aquaculture and tourism

14.A Increase scientific knowledge, develop research capacity and transfer marine technology, taking into account the Intergovernmental Oceanographic Commission Criteria and Guidelines on the Transfer of Marine Technology, in order to improve ocean health and to enhance the contribution of marine biodiversity to the development of developing countries, in particular small island developing States and least developed countries

14.B Provide access for small-scale artisanal fishers to marine resources and markets

14.C Enhance the conservation and sustainable use of oceans and their resources by implementing international law as reflected in UNCLOS, which provides the legal framework for the conservation and sustainable use of oceans and their resources, as recalled in paragraph 158 of The Future We Want

Sustainable Development Goal 15

Protect, restore and promote sustainable use of terrestrial ecosystems, sustainably manage forests, combat desertification, and halt and reverse land degradation and halt biodiversity loss

Preserving diverse forms of life on land requires targeted efforts to protect, restore and promote the conservation and sustainable use of terrestrial and other ecosystems. Goal 15 focuses specifically on managing forests sustainably, halting and reversing land and natural habitat degradation, successfully combating desertification and stopping biodiversity loss.

Infrastructure can play a role in promoting sustainable land and forest management practices in order to combat desertification, soil degradation, and biodiversity loss in a range of biomes, including wetlands, mountains, forests and drylands. Appropriate flood risk protection can reduce the risk of land degradation, while the electrification of communities through renewable generation can provide a sustainable energy source that does not rely on the use of solid biomass fuels and can thus reduce deforestation. The planning of transport infrastructure such as roads can be designed to minimise impacts on terrestrial ecosystems, such as habitat destruction and fragmentation. The provision of infrastructure to provide suitable management of waste and wastewater can eliminate contamination or pollution. Using digital technology, remote monitoring using electronic tags can provide necessary data and information to combat poaching and the lucrative trade in illegal wildlife parts.

TARGETS

15.1 By 2020, ensure the conservation, restoration and sustainable use of terrestrial and inland freshwater ecosystems and their services, in particular forests, wetlands, mountains and drylands, in line with obligations under international agreements

15.2 By 2020, promote the implementation of sustainable management of all types of forests, halt deforestation, restore degraded forests and substantially increase afforestation and reforestation globally

15.3 By 2030, combat desertification, restore degraded land and soil, including land affected by desertification, drought and floods, and strive to achieve a land degradation-neutral world

15.4 By 2030, ensure the conservation of mountain ecosystems, including their biodiversity, in order to enhance their capacity to provide benefits that are essential for sustainable development

15.5 Take urgent and significant action to reduce the degradation of natural habitats, halt the loss of biodiversity and, by 2020, protect and prevent the extinction of threatened species

15.6 Promote fair and equitable sharing of the benefits arising from the utilization of genetic resources and promote appropriate access to such resources, as internationally agreed

15.7 Take urgent action to end poaching and trafficking of protected species of flora and fauna and address both demand and supply of illegal wildlife products

15.8 By 2020, introduce measures to prevent the introduction and significantly reduce the impact of invasive alien species on land and water ecosystems and control or eradicate the priority species

15.9 By 2020, integrate ecosystem and biodiversity values into national and local planning, development processes, poverty reduction strategies and accounts

15.A Mobilize and significantly increase financial resources from all sources to conserve and sustainably use biodiversity and ecosystems

15.B Mobilize significant resources from all sources and at all levels to finance sustainable forest management and provide adequate incentives to developing countries to advance such management, including for conservation and reforestation

15.C Enhance global support for efforts to combat poaching and trafficking of protected species, including by increasing the capacity of local communities to pursue sustainable livelihood opportunities

Sustainable Development Goal 16

Promote peaceful and inclusive societies for sustainable development, provide access to justice for all and build effective, accountable, and inclusive institutions at all levels

The 2030 Agenda for Sustainable Development aims to promote peaceful and inclusive societies based on respect for human rights, the rule of law and transparent, effective and accountable institutions. The elimination of violence and armed conflict should be accompanied by strong institutional development at all levels of government, as well as universal access to justice, information and other fundamental freedoms.

Targeted infrastructure interventions can contribute to reductions in violence, conflict and crime across societies. For example, widespread electrification may enable steps to improve safety in communities such as the deterrent of violence through street lighting. Efficient roads and waterways can allow law enforcement timely access to communities in order to respond to emergencies and crimes. While digital technology can enable more sophisticated crimes, including cybercrime and illicit financial flows, it also provides the necessary capabilities to combat them. In some regions, infrastructure may have been damaged or destroyed by war or conflict; rebuilding it is a key step to restoring peace and order to the affected communities. The construction of police stations, courts and prisons to adequate standards will improve law enforcement and access to justice.

However, equally important to the achievement of peaceful and inclusive societies are strong and effective institutions. Participatory decision-making will require people, including the most vulnerable, to have access to institutions of governance at all levels. Such access may require better transport links – for example, to ensure all citizens can express their right to vote. Access and accountability can also be improved with the help of digital communications infrastructure, by allowing wider dissemination of information and online forums for public consultation.

TARGETS

16.1 Significantly reduce all forms of violence and related death rates everywhere

16.2 End abuse, exploitation, trafficking and all forms of violence against and torture of children

16.3 Promote the rule of law at the national and international levels and ensure equal access to justice for all

16.4 By 2030, significantly reduce illicit financial and arms flows, strengthen the recovery and return of stolen assets and combat all forms of organized crime

16.5 Substantially reduce corruption and bribery in all their forms

16.6 Develop effective, accountable and transparent institutions at all levels

16.7 Ensure responsive, inclusive, participatory and representative decision-making at all levels

16.8 Broaden and strengthen the participation of developing countries in the institutions of global governance

16.9 By 2030, provide legal identity for all, including birth registration

16.10 Ensure public access to information and protect fundamental freedoms, in accordance with national legislation and international agreements

16.A Strengthen relevant national institutions, including through international cooperation, for building capacity at all levels, in particular in developing countries, to prevent violence and combat terrorism and crime

16.B Promote and enforce non-discriminatory laws and policies for sustainable development

Sustainable Development Goal 17

Strengthen the means of implementation and revitalize the global partnership for sustainable development

Achieving the ambitious targets of the 2030 Agenda requires a revitalized and enhanced global partnership that brings together Governments, civil society, the private sector, the United Nations system and other actors, mobilizing all available resources. Strengthening partnerships for the SDGs at the national or international scale requires infrastructure that can facilitate integration through information sharing, knowledge transfer and capacity building incorporating a variety of infrastructure types, including governance, educational and economic infrastructure. The targets for this Goal can also be strongly supported by transportation and digital communications networks that can bring actors together and capitalise on global partnerships to most effectively implement sustainable development policies.

At a national level, mobilising resources for developing countries can benefit from digital technology, such as simplified tax collection through electronic channels that allow governments to raise the funds necessary to implement sustainable development targets. Increasing trade integration through access to international markets can be enhanced by the construction of air and seaports and the road and rail networks that connect them to local producers. Such access infrastructure may overcome particular challenges in landlocked countries, small islands, or rural regions. Through the sharing of technologies, in addition to technical skills training outlined in Goal 4, countries can increase the value added of local production while developing and implementing environmentally sustainable technologies suited to local contexts.

TARGETS

FINANCE

17.1 Strengthen domestic resource mobilization, including through international support to developing countries, to improve domestic capacity for tax and other revenue collection

17.2 Developed countries to implement fully their official development assistance commitments, including the commitment by many developed countries to achieve the target of 0.7 per cent of ODA/GNI to developing countries and 0.15 to 0.20 per cent of ODA/GNI to least developed countries ODA providers are encouraged to consider setting a target to provide at least 0.20 per cent of ODA/GNI to least developed countries

17.3 Mobilize additional financial resources for developing countries from multiple sources

17.4 Assist developing countries in attaining long-term debt sustainability through coordinated policies aimed at fostering debt financing, debt relief and debt restructuring, as appropriate, and address the external debt of highly indebted poor countries to reduce debt distress

17.5 Adopt and implement investment promotion regimes for least developed countries

TECHNOLOGY

17.6 Enhance North-South, South-South and triangular regional and international cooperation on and access to science, technology and innovation and enhance knowledge sharing on mutually agreed terms, including through improved coordination among existing mechanisms, in particular at the United Nations level, and through a global technology facilitation mechanism

17.7 Promote the development, transfer, dissemination and diffusion of environmentally sound technologies to developing countries on favourable terms, including on concessional and preferential terms, as mutually agreed

17.8 Fully operationalize the technology bank and science, technology and innovation capacity-building mechanism for least developed countries by 2017 and enhance the use of enabling technology, in particular information and communications technology

CAPACITY BUILDING

17.9 Enhance international support for implementing effective and targeted capacity-building in developing countries to support national plans to implement all the sustainable development goals, including through North-South, South-South and triangular cooperation

TRADE

17.10 Promote a universal, rules-based, open, non-discriminatory and equitable multilateral trading system under the World Trade Organization, including through the conclusion of negotiations under its Doha Development Agenda

17.11 Significantly increase the exports of developing countries, in particular with a view to doubling the least developed countries' share of global exports by 2020

17.12 Realize timely implementation of duty-free and quota-free market access on a lasting basis for all least developed countries, consistent with World Trade Organization

decisions, including by ensuring that preferential rules of origin applicable to imports from least developed countries are transparent and simple, and contribute to facilitating market access

SYSTEMIC ISSUES

POLICY AND INSTITUTIONAL COHERENCE

17.13 Enhance global macroeconomic stability, including through policy coordination and policy coherence

17.14 Enhance policy coherence for sustainable development

17.15 Respect each country's policy space and leadership to establish and implement policies for poverty eradication and sustainable development

MULTI-STAKEHOLDER PARTNERSHIPS

17.16 Enhance the global partnership for sustainable development, complemented by multi-stakeholder partnerships that mobilize and share knowledge, expertise, technology and financial resources, to support the achievement of the sustainable development goals in all countries, in particular developing countries

17.17 Encourage and promote effective public, public-private and civil society partnerships, building on the experience and resourcing strategies of partnerships

DATA, MONITORING AND ACCOUNTABILITY

17.18 By 2020, enhance capacity-building support to developing countries, including for least developed countries and small island developing States, to increase significantly the availability of high-quality, timely and reliable data disaggregated by income, gender, age, race, ethnicity, migratory status, disability, geographic location and other characteristics relevant in national contexts

17.19 By 2030, build on existing initiatives to develop measurements of progress on sustainable development that complement gross domestic product, and support statistical capacity-building in developing countries

ICE timeline

1818 ICE was founded

1820 Thomas Telford became the first president of ICE

1826 Menai Straits Bridge opened

1828 ICE achieves Royal Charter

1830 Liverpool and Manchester Railway opened

1843 Thames Tunnel opened

1863 Metropolitan Railway opened

1866 Bazalgette's London Sewers completed

1894 Manchester Ship Canal opened

1946 Heathrow Airport opened for civil aviation

1958 Preston bypass became the UK's first stretch of motorway

1973 Sydney Opera House opened

1975 Royal Charter updated by Queen Elizabeth II

1984 Thames Barrier opened

1994 Channel Tunnel opened

2012 Queen Elizabeth Olympic Park opened

2016 Infrastructure Learning Hub (ILH) and Members' Learning Hub (MLH) opened

2018 Bicentenary of ICE

2023 Thames Tideway Tunnel due for completion

About ICE

The Institution of Civil Engineers is one of the world's most respected professional engineering associations. Since we were formed 200 years ago, we have attracted some of the most famous and influential civil engineers in history.

In 1818 three young engineers met in a London coffee shop and founded the Institution of Civil Engineers (ICE), the world's first professional engineering body.

They had hoped that lots of engineers from different engineering backgrounds would join the institution. However, civil engineering hadn't really become an official profession yet, and before the 18th century most engineers were in the armed forces.

After two years of struggling to attract new members, ICE asked Thomas Telford (right) to become its first President. His appointment in 1820 not only gave ICE a major boost, it also played a huge part in shaping who we are today.

Telford designed and built all types of infrastructure; from churches to castles, canals to harbours, tunnels to bridges. He was also given the nickname 'the Colossus of Roads' because his designs were used to construct all major British highways (during his lifetime he built more than 1,000 miles of roads).

Using his political and social connections, Telford helped to bring in many new members, from the UK and overseas. But his most important role was getting ICE's Royal Charter in 1828. Our charter (updated by Queen Elizabeth II in 1975), gives us our status as the leading institution for the civil engineering profession.

Since our small beginnings, ICE has become home to many of history's greatest engineers as past presidents and members, and 200 years later, we have more than 95,000 members.

About the publisher

Artifice Press is one of the world's leading publishers of high-quality books on engineering, architecture, urbanism and the built environment. Documenting spectacular projects, from civil engineering masterpieces to the work of award-winning architects, it works collaboratively with independent studios, international institutions, academics and architectural practices to create hardback books that are engaging, insightful and beautiful.

UN75: Sustainable Engineering In Action is the second publication in a series of five collaborations between Artifice Press and ICE. In 2018, Artifice Press published *Shaping The World: Two Hundred Years of the Institution of Civil Engineers*, which celebrated the Institution's bicentenary by profiling 200 of the greatest civil engineering feats of the past two centuries. Writing in *The Times*, Matthew Parris described it as "unputdownable were it not for the difficulty of picking it up" and "a thousand times more gripping than its title".

"Put it on your coffee table and your coffee will grow cold before you look up from its 400 pages," wrote Parris. "The editors have selected hundreds of the world's most astonishing schemes — bridges, tunnels, dams, palatial sewers, airports, grand canals, giant turbines and impossible railways, each with stunning photographs and a few hundred words of clear-to-a-layman explanatory text."

Other prominent Artifice Press partnerships include working with the Royal Institute of British Architects (RIBA), having sponsored the 2018 RIBA Building Of The Year Awards and published an accompanying book, which examined all 49 award-winning structures.

Recent Artifice Press titles exploring iconic building projects include: *70 St Mary Axe*, which charts the creation of Foggo Associates' innovative and distinctive "Can of Ham" building in the City of London, from concept to completion; *Thinking Outside The Box: Reimagining Television Centre*, which explores how the BBC's iconic former home in White City has been transformed over the past two decades; and *The Evolution Of A Building Complex* – Jeffry Kieffer's study of the Salk Institute complex in California and the work of its architect, Louis I Kahn.

Further publications include monographs for renowned practices such as Eric Parry Architects in London, SLCE Architects in New York and DP Architects in Singapore. Earlier this year, Artifice Press published *From Now On*, a study on the RIBA award-winning practice Hollaway, and *Space For Architecture*, a book about the Anglo-Irish practice O'Donnell + Tuomey Architects, creators of the Photographers' Gallery and the LSE's Saw Swee Hock Student Centre.

Artifice Press is part of the SJH Group, a creative multimedia organisation that prints around 300,000 books each year. "Artifice Press specialises in impeccably designed publications that transcend disciplinary constraints," says Publishing Director Anna Danby. "We create beautiful hardback books featuring stunning photography and the highest possible production standards. But we're not afraid to engage with the key debates that surround the built environment, addressing the challenges of urbanism, climate, education and technology. These are books with beauty and brains."
www.artificebooksonline.com

Acknowledgments

Contributors

Charles Jensen, Commissioning Editor, ICE
Anna Plowdowski, Copy Editor, ICE

John Lewis pp. 16–27
David Balmforth, Rob Curd pp. 34–36
UNOPS pp. 37–40
International Labour Organization pp. 42–45
UNOPS pp. 46–48
Rebecca Lovelace pp. 49–51
Alexander H. Hay pp. 52–55
Ian Bingham, Lance Foxall pp. 56–57
World Health Organization pp. 68–69
World Food Programme pp. 70–73
UNOPS pp. 74–77
Andrew Swan, Pete Skipworth pp. 78–79
Edoardo Borgomeo pp. 80–81
Management Team of Ashghal's DSSIP Project pp. 82–83
NMCN and Sheffield City Council pp. 84–85
Aristotles Tegos, Aristotelis Koskinas pp. 86–87
Mohammed Safaraz Gani Adnan pp. 88–89
Bill Douglas, Alan McGowan pp. 90–91
Daniel Whiteley pp. 92–93
WaterAid pp. 94–97
UNESCO pp. 118–21
Shamsul Hadi Shams, UNITAR pp. 122–24
Nidhi Nagabhatla, UN University pp. 125–27
UNICEF pp. 128–31
UN Women pp. 132–33
UNHCR, The UN Refugee Agency pp. 134–35
UNOPS pp. 136–38
UNOPS pp. 139–41
UNOPS pp. 142–45
Kirsten MacAskill pp. 146–47

Brian Reed pp. 148–49
Darren White, Monica Lobo pp. 150–51
Jo de Serrano pp. 152–54
UNOPS pp. 168–70
Neil Strong, Rossa Donovan pp. 171–73
Daniel Adshead pp. 174–75
Lena Fuldauer pp. 176–77
Patrick James, AbuBakr Bahaj pp. 178–79
Jenny Cooke, Network Rail pp. 180–81
Colin M. Rose and Julia A. Stegemann pp. 182–83
BRE pp. 184–85
BRE pp. 186–87
Daniel Whiteley pp. 188–89
Rob Paris pp. 190–91
Daniel Whiteley pp. 192–94
UN-Habitat pp. 214–15
UNOPS pp. 216–18
Bill Addis pp. 219–21
Gabriele Manoli, Anthony J. Parolari, Athanasios Paschalis, Edoardo Daly, Simone Fatichi pp. 222–23
UWE pp. 224–25
Patrick James, AbuBakr Bahaj, Victor Atehortua pp. 226–27
BRE pp. 228–29
Ed Cole and David Stewart, Jacobs pp. 230–31
BRE pp. 232–33
Transport Scotland, BRE pp. 234–35
Network Rail, BRE pp. 236–37
Daniel Whiteley pp. 238–40